T0302283

Robot Ethics and the Innovation Economy

This book provides an authoritative resource on the topic of intelligent robots, artificial intelligence and the ethical implications of these revolutionary innovations. It examines the moral and ethical problems that arise in relation to the development, design and use of intelligent robots, which are capable of autonomous or semi-autonomous decision-making. These problems might relate, for example, to medical robots, driverless cars, intelligent military drones, pedagogical robots, police robots, legal robots and many others.

The main question addressed in this book is how we can understand, explain and apply the concept of ethics in relation to intelligent robots and artificial intelligence. In each chapter, the author examines a different aspect of this question. The author also questions how we can ensure that intelligent robots are of service to humans and under what conditions intelligent robots could become more ethical than humans. The book employs an original approach to examining this cutting-edge research question, combining different research areas, and offers a wealth of practical relevance and real-world examples, illustrated through vivid case studies. With its jargon free approach and a dedicated chapter on relevant concepts at the end, this book is also accessible to readers without prior knowledge on intelligent robots and the Fourth Industrial Revolution.

By providing a general account of this debate, and of the consequences of the innovations resulting from these trends, the book serves as an important contribution to the discussion and will find a natural readership among scholars and students of the innovation economy and those concerned with the ethical considerations arising in the wake of the Fourth Industrial Revolution.

Jon-Arild Johannessen is a Professor (full) in Leadership at Kristiania University College, Oslo, Norway. He holds a Master of Science from Oslo University in History, and a Ph.D. from Stockholm University in Systemic Thinking. Previously, he has been professor (full) in Innovation, at Syd-Danske University, Denmark, and in Management at The Arctic University, Norway. At Bodø Graduate School of Business, Norway, he had a professorship (full) in Information Management. At Norwegian School of Management (BI) he has been professor (full) in Knowledge Management.

Routledge Studies in the Economics of Innovation

The Routledge Studies in the Economics of Innovation series is our home for comprehensive yet accessible texts on the current thinking in the field. These cutting-edge, upper-level scholarly studies and edited collections bring together robust theories from a wide range of individual disciplines and provide in-depth studies of existing and emerging approaches to innovation, and the implications of such for the global economy.

The Impact of the Sharing Economy on Business and Society
Digital Transformation and the Rise of Platform Businesses
Edited by Abbas Strømmen-Bakhtiar and Evgueni Vinogradov

Automation, Innovation and Work
The Impact of Technological, Economic, and Social Singularity
Jon-Arild Johannessen and Helene Sætersdal

The Political Economy of Digital Automation
Measuring its Impact on Productivity, Economic Growth and Consumption
Sreenath Majumder and Anuradha SenGupta

Artificial Intelligence, Automation and the Future of Competence at Work
Jon-Arild Johannessen

Capitalism, Power and Innovation
Intellectual Monopoly Capitalism Uncovered
Cecilia Rikap

Robot Ethics and the Innovation Economy
Jon-Arild Johannessen

For more information about this series, please visit: www.routledge.com/ Routledge-Studies-in-the-Economics-of-Innovation/book-series/ECONINN

Robot Ethics and the Innovation Economy

Jon-Arild Johannessen

Routledge
Taylor & Francis Group

LONDON AND NEW YORK

First published 2021
by Routledge
2 Park Square, Milton Park, Abingdon, Oxon OX14 4RN

and by Routledge
605 Third Avenue, New York, NY 10158

Routledge is an imprint of the Taylor & Francis Group, an informa business

© 2021 Jon-Arild Johannessen

British Library Cataloguing-in-Publication Data
A catalogue record for this book is available from the British Library

Library of Congress Cataloging-in-Publication Data
Names: Johannessen, Jon-Arild, author.
Title: Robot ethics and the innovation economy / Jon-Arild Johannessen.
Description: Abingdon, Oxon; New York, NY: Routledge, 2021. |
Series: Routledge studies in the economics of innovation |
Includes bibliographical references and index.
Identifiers: LCCN 2020056656 (print) | LCCN 2020056657 (ebook)
Subjects: LCSH: Robotics–Moral and ethical aspects. |
Robotics–Economic aspects. | Computer-human interaction.
Classification: LCC TJ211.28 .J64 2021 (print) |
LCC TJ211.28 (ebook) | DDC 174/.9629892–dc23
LC record available at https://lccn.loc.gov/2020056656
LC ebook record available at https://lccn.loc.gov/2020056657

ISBN: 978-1-032-00513-3 (hbk)
ISBN: 978-1-032-00512-6 (pbk)
ISBN: 978-1-003-17449-3 (ebk)

Typeset in Bembo
by Newgen Publishing UK

Contents

Figures

Foreword

This book is intended as an authoritative resource on the topics of knowledge processes, intelligent robots, artificial intelligence and ethics. These topics are very widely debated in scholarly journals, newspapers and online. By providing a general account of this debate, and of the consequences of the innovations resulting from these trends, we believe that we have contributed to the debate about the ethical implications of intelligent robots.

We have not assumed any particular knowledge of intelligent robots and the Fourth Industrial Revolution on the part of the reader. Accordingly, we have done our best to avoid technical jargon and other specialized academic content. We have tried to make the text understandable by the general reader, so that an alert student will be easily able to understand the individual chapters of this book. The use of some technical words and expressions has been unavoidable, however. Accordingly, we have included a dedicated chapter on concepts at the end of the book. The reader can refer to this at any time to find out about the meanings of different concepts and perhaps also about their origins and development.

The main question addressed in this book is: how can we understand, explain and apply ideas about ethics related to intelligent robots and artificial intelligence? In each chapter of this book, we examine a different aspect of this question. Throughout the book we refer to practical examples (case studies), theoretical points and the practical utility value of the questions we are investigating for individuals, organizations and society.

1 Intelligent robots and ethics

The key ideas of the chapter

1. Intelligent robots will be increasingly able to reflect on ethics, and will be more able to act ethically than humans.
2. Responsibility for a mistake made by a military robot, for example an incident where civilians, sick people or children are injured or killed, lies with the drone pilot who is operating the robot. An analogous situation would be that of a car driver in a situation where the car brakes fail.
3. Through the development of the intelligent robot, we have become what we actually are: rational, logical, instrumental actors who value what can be weighed, measured and counted. We have created intelligent robots in our own image, and we are frightened by what we see.
4. The ethical mirror: fear of intelligent robots is nothing more than fear of what we are turning into as humans: rational, logical agents who do not take into account the intelligence of a smile, emotions or the touch of a hand.
5. In intelligent robots, ethical programming is structured hierarchically as three levels of logic: the contextual and cultural level; the situational level; and the operational level.
6. Intelligent robots reflect on unethical actions performed by unintelligent humans.
7. Under certain conditions, intelligent robots can have autonomous moral responsibility. This is an innovation in the context of moral philosophy.

Introduction

This book is about the moral and ethical problems that arise in relation to the development, design and use of intelligent robots, which are capable of autonomous or semi-autonomous decision-making. These problems might relate, for example, to medical robots, driverless cars, intelligent military drones, pedagogical robots, police robots, legal robots and so on.

Robot ethics is a branch of applied ethics, which examines the ethical problems that arise when intelligent robots are designed and applied in practice.

The field of robot ethics is also referred to as roboethics (Veruggio, 2005: 1–4; Operto, 2011). Ethics concerns whether actions are right or wrong.

How can we ensure that intelligent robots are of service to humans? Under what conditions could intelligent robots be more ethical than humans? These are the questions that we examine in this book.

Ethics is a field with many theories and perspectives. First, there are meta-ethics, normative ethics and applied ethics. In this book, we are concerned with applied ethics. Second, there are various ethical theories, including virtue ethics, deontological theory, utilitarian theory, justice-and-fairness theory, egoism theory, value-based theory and case-based theory, to name just the most well-known. In this book, our discussions are based on a systemic perspective on ethics. We have described and analysed this perspective in Appendix 1 and Appendix 2.

Can intelligent robots learn to act in a moral manner? Are intelligent robots responsible for their actions? Under what conditions can intelligent robots perform moral actions? Can intelligent robots have a sense of morality? Can robots be viewed as moral agents? Much attention has been devoted recently to these questions (Rodogno, 2016; Coeckelbergh, 2014; Gunkel, 2014). According to these authors, the answer is no. In other words, robots cannot be viewed as moral agents. A moral agent is considered to have the capacity to reflect, evaluate, make rational choices, decide and act in a given situation (Cave, 2002: 4). Philosophers also suggest that free will is a criterion for moral agency (Shaun & Knobe, 2007: 663–685). If you have moral responsibility, you are a moral agent, and vice versa.

Moral responsibility is not necessarily linked to legal responsibility. We can assume that in the near future, intelligent robots and informats[1] will be capable of being considered to be moral agents in accordance with the above definition. So does this mean that they are also responsible for their actions? We would probably all consider it absurd for an intelligent robot to be prosecuted and punished for its actions.

If we frame the question differently, however, we get a different perspective. Can robots perform moral actions? When considering this question, we are only interested in whether robots can make decisions based on what action would be morally right or wrong in a given situation. We are talking here about intelligent robots and informats. Although intelligent robots and informats can perform moral actions, it would be meaningless to say that robots are morally responsible (Asimov, 2008). It is equally meaningless to talk about robots having rights on a par with humans. In the not-so-distant future, we can envisage some humans having some of their organs replaced with nanorobots. But we can also imagine humans having other types of nanorobots added to their bodies in order to enhance their performance in some way or other. When humans are the starting point, then the answer is that they are both moral agents and responsible for their actions, regardless of how much technology is implanted in their bodies. Obviously, electronic linking structures are not moral per se. They do become part of a moral system, however, if they are structurally linked to a human being (Bunge, 2013).

We encounter somewhat similar problems in relation to tools used in genome engineering and social robots used in the healthcare sector (Coeckelbergh, 2010; De Grey, 2013). These tools can make major changes to our "natural" genetic composition and can change human performance and pathologies (Doudna & Sternberg, 2018).

Acting in a way that is morally wrong involves the feelings and emotions of others, at one level or another (Rodogno, 2016: 42). Accordingly, a simple definition of morally wrong behaviour could be that one is oppressing other individuals, in one way or another. We could simplify the definition of morally correct behaviour to say that it means showing respect for others, taking responsibility for others and treating others in a dignified fashion (RRD)[2] (Benhabib, 2004). From this perspective, we can say that robots and informats can act in ways that are morally right and wrong. This becomes clear if we envisage intelligent drones as "killing machines".

If a robot is designed so that it can act rationally, and all feelings and emotions are designed out of the programs that control it, then the robot is approaching the classic definition of a psychopath – a person who is highly rational but lacks normal emotions (Ronson, 2011). It is a long road from the Turing test of the 1950s, which required a human interacting with a robot to believe that he or she was interacting with a human, to reaching this technology singularity (Shanahan, 2015). In the latter case, robots will have super-intelligence. They will design their own code through a process of learning. It seems reasonable to assume that artificial intelligence, robots and informats will have decisive significance, both for moral actions and for how they affect society. The decision to design artificial intelligence, social robots, informats, medical robots, military robots and so on is an ethical choice. When this technology attains technological singularity, it will also perform moral actions in the context of RRD (respect, responsibility, dignity). Just as with the climate crisis, it is imperative that we act now to impose limits on what is permissible in the development of artificial intelligence. Just as with the climate crisis, it will be too late when we have reached the point of no return. Technological singularity is expected to occur in around 2040.[3] The period from today to 2040 is our window of opportunity. After 2040, it may be too late (Shanahan, 2015).[4]

This chapter is intended to provide an introduction to the central concepts discussed in this book. It is also intended to provide a general overview of the main question and the subsidiary research questions that we examine in this book. In this introductory chapter, we will touch in a general way on each of the chapters in the book, so that readers gain an idea of what they will encounter in the respective chapters later on in the book, where the questions are considered in more detail. In other words, Chapter 1 not only summarizes what this book is about, but is also a free-standing investigation of the questions listed below.

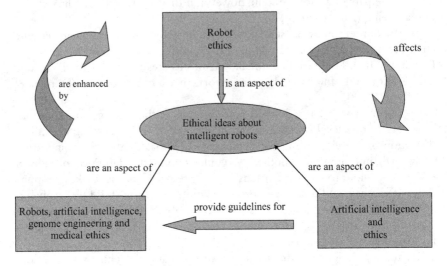

Figure 1.1 Innovation and ethics.

The question we examine in this chapter is as follows: how can we understand, explain and apply ideas about ethics to intelligent robots and artificial intelligence?

In order to tackle this general question, we have developed three subsidiary research questions:

1.   What is robot ethics?
2.   How can we reflect on artificial intelligence and ethics?
3.   What ethical problems will be caused by the application of intelligent robots, artificial intelligence and genome editing in the healthcare sector?

We have summarized the introduction in Figure 1.1, which also illustrates how we have structured this chapter, as well as showing how this book is structured overall.

Robot ethics

The question we investigate here is as follows: what is robot ethics?

To paraphrase a quote from Aristotle, one could say that the best way to learn about acting ethically is to make it one's everyday practice (Tzafestas, 2016: 1). One could imagine this being embedded into the algorithms of intelligent robots. A basic ethical algorithm in all intelligent robots could be something like this: if an ethical situation arises, then always choose the behaviour that is considered ethically good. The meaning of ethically good behaviour in different contexts would then have to be programmed into the robot like a meta-algorithm.

Robot ethics relates directly to the ethics of technology and the ethics of innovation (Allen et al., 2006: 12–17; Hall, 2000: 28–46). Robot ethics is linked to the development, design and application of intelligent robots in society. As a general rule, robot ethics is concerned with autonomous robots, i.e., robots that are not linked to a system (Tzafestas, 2016: 2). It is our opinion here that having intelligent robots linked up through a global artificial intelligence network would be more expedient. The reason for this is that intelligent robots would then be able to learn more easily from other robots through a process of trial and error, as well as from other robots' behaviours and responses in different situations.

Is it ethically correct to program ethical codes into intelligent robots? Surely, ethics is only concerned with the human domain? Let us imagine a situation where an intelligent robot has ethical codes written into its program such as a semi-automated car. If the car drives into someone on a crowded street, who is then responsible for the injury suffered by the accident victim? Is it the robot, the designer, the "driver" of the car or the owner of the car? Can a robot be held responsible for its "actions"? If so, then the robot will be summoned to court, be judged and possibly sentenced. Of course, this sounds completely absurd – that a car would be summoned to court to face justice! What about the person who designed the algorithm installed in the vehicular robot? Is that person responsible for the "actions" of the robot? Or is it the business for whom the designer works that is responsible? We see the questions starting to mount up – questions to which we today have no clear answers. The most obvious answer is that the person who owns and "drives" the car is the one who will most probably be held responsible.

To simplify the problem, let us assume that it is the car's owner who is "driving" the car. In such a case, it is the driver who is responsible at all times for the injuries inflicted on third parties, even if the intelligent robot took over control of the car in a critical situation and acted "ethically". It is quite possible that the company that designed the algorithm will be the one judged guilty in a court trial, but in the first instance, the responsibility rests on the "driver". In this case, we have considered that it will be the "driver" of the car who will initially be held responsible, and then possibly the company.

In other technologies, it may be more complicated to decide who should be held responsible in the case of an accident or injury. Suppose an intelligent robot diagnoses and performs an operation on a patient. Who will be held responsible if the diagnosis is incorrect and the patient is permanently injured by the operation? In the first instance, we assume that it is the doctor who assists the robot who is responsible. In such a case, similar to the driver of the car, it is the doctor who will be held responsible for the operation that resulted in a permanent injury to the patient. Ultimately, it will be the hospital that is the legal subject and will be held responsible in a court case. However, let us imagine a future where there is no doctor to assist the intelligent robot. Who is then responsible for the errors that may arise? In the same way as in the example of the car, we believe that it is the organization, i.e., the hospital, that will ultimately be held responsible for the errors committed by the robot.

In the future, it will be difficult to define ethical principles across cultures for the design of intelligent robots (Wallach & Allen, 2009). Nevertheless, there seem to be two main perspectives that will be relevant across cultures. First, there is a theoretical starting point, and second, a practical-empirical point of departure. The theoretical starting point will concern some general ethical rules, which will be installed in robots. The practical perspective will be evolutionary and lead to ethical rules being evolved, as practical experience is gained from using the robots.

In this book, we are more oriented towards a third variant: the interaction between theory and practice. First, some general ethical rules will be installed in the intelligent robots. Second, a learning system will be installed, where feedback from practice will be given priority over the theoretical ethical rules.

Let us consider a scenario where general ethical rules may be challenged. We can imagine that a general ethical rule, say Ethical Rule 1, stating that no intelligent robot is permitted to harm a human being, is installed in the robots. However, would this rule be applicable in all situations? What if the intelligent robot in question is given the task of preventing a massive terrorist attack on a government building, where it is likely that hundreds of civilians and officials will be killed? If the robot is an intelligent drone that has the task of eliminating the terrorist in order to prevent the terrorist attack, will Ethical Rule 1 still apply? Or will Ethical Rule 1 be set aside in such exceptional situations? It may further be imagined that the robot is designed so that it obeys any command from any human being. What will happen to such an ethical rule if a human being, who happened to be a co-conspirator of the terrorist, gave the command not to eliminate the terrorist? Wouldn't it be more ethically correct if the robot did not obey the command and eliminated the terrorist? Consequently, a general rule, of the type "intelligent robots must obey any human", can result in major ethical problems, while material, economic and legal problems may also occur.

For instance, installing a general rule: "you must not kill a human being", as in the example mentioned above, would be both wrong and immoral, because killing one person, such as a terrorist, could prevent hundreds, perhaps even thousands, of civilians from being injured and killed. We are therefore of the opinion that the most appropriate action should take into consideration the context, situation and operational rules.

In the same way as it is problematic to establish a general rule, as shown above, it will also be problematic to rely exclusively on the type of modern deontological (duty-based) ethical rules proposed by Gert (1988) and Gips (1992): don't kill; don't cause pain; don't disable; don't deprive of freedom; don't deprive of pleasure; don't deceive; keep your promise and don't cheat. None of the above eight ethical rules would be valid with regard to the appropriate course of action to be taken by an intelligent robot to prevent an imminent terrorist attack aimed at killing all the members of a government. Consequently, installing only a duty-based ethics management system in an intelligent robot could result in major problems. In our opinion, it is also problematic if humans exclusively follow a duty-based ethical code.

Consequentialism is the approach to ethics which sees morality as concerning the right overall consequences of an action. Although intentions may be good, it is the consequences against which the action is judged. This view is exemplified by the saying: "the road to hell is paved with good intentions".[5] According to consequentialism, the best moral action is the one that leads to the best future moral consequence. This applies to humans, and it is probable that it will apply to intelligent robots in the future.

From a theoretical perspective, the intelligent robot should be able to develop a conceptual model of the present situation and then act accordingly. Consequently, it does not need to have an overview of all the conceivable situations, but instead needs to be able to develop a model of the present situation, and be able to "see" the consequences of the actions it does or does not carry out.

One may also imagine that the intelligent robot will act on the basis of empirically based situations, examples, cases and the like, that have been installed in its memory. When the robot has gained more experience and learned to build new examples into its memory, it will be able to cope with situations that the designer had not considered when the robot was first programmed. In such situations, we say that the intelligent robot bases its actions on empirical generalizations.

Both in terms of conceptual generalization and empirical generalization, the intelligent robot will be able to develop alternative action strategies, i.e., it will be able to envision (calculate) the different consequences that will result from different actions, based on different situations. We can further assume that the intelligent robot will also be able to envision the different situations and consequences of its actions, based on what is the right action/not the right action and what is morally good or bad.

In the future, robot ethics will be concerned with at least two categories of intelligent robots. First, there will be those robots that are autonomous and independent of human control, and second, those that are controlled directly and indirectly by the person(s) using them. Robot ethics will mainly focus on the autonomous robots, which will make decisions across, and in some cases despite, human involvement. The simplest example is an intelligent robot in a semi-autonomous car, where the driver controls the car, but where the intelligent robot intervenes in crisis situations where, for example, a choice has to be made between colliding with an old man or a child. This example could be further developed by assuming that the little child was unknown to the driver, whereas the old man was the driver's father. The intelligent robot may not discriminate in the same way as the driver and may make different choices.

Based on the above description, there are some problems that will have to be addressed regarding robot ethics. First, the nature of the communication channels between man and robot will have to be considered. Second, the problems concerning the system for overriding a robot will also need to be

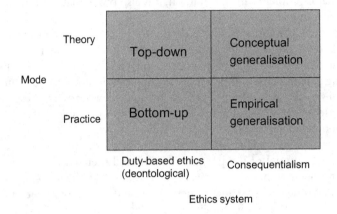

Figure 1.2 Aspects of robot ethics: A typology.

examined. For instance, it is not at all certain that it would be ethically correct in all situations for humans to override the decision made by an intelligent robot, cf. the example above, where the driver of the car might choose to override the decision of the robot, in order to avoid injuring his own father. Third, the placement of the ethical function in the robot's decision-making structure will have to be considered. For instance, should it be placed above or below the utility function? In practice, this will concern whether priority should be given to the ethical or the utility aspect of a situation. Consequently, when designing a robot, it is important to grasp the problems of the decision-making architecture.

In Figure 1.2 we have developed a typology that provides a tentative answer to the question concerning aspects of robot ethics.

AI and ethics

In this section, we will examine the following question: what is the nature of the problems concerning the ethics of artificial intelligence (AI)?

It is difficult to say with certainty whether intelligent robots can be designed to act ethically. On the other hand, it is safe to say that humans also struggle with the same "design error", that is, many intelligent people act unethically, even though they know what is ethically right. Thus, the debate concerning artificial intelligence will not only be of benefit when designing intelligent robots, but will perhaps also contribute to raising awareness of ethical problems among the general population.

Artificial intelligence is not a new invention; its history goes back more than 60 years to 1956, when the Dartmouth Summer Research Project on AI took place, where McCarthy established the concept of artificial intelligence (McCarthy & Hayes, 1969: 463–502). The development of artificial intelligence was linked to the idea of developing machines that could think and change themselves after making errors (Dreyfus, 1979). Whether artificial intelligence can be anything other than goal-seeking is discussed among computer scientists. Dreyfus (1979) takes the position that intelligent machines can only be goal-seeking, while what makes us human, says Dreyfus, is that we are value-seeking in addition to goal-seeking. On the other hand, researchers such as Alan Turing and Herbert Simon claim that artificial intelligence can be developed to incorporate both goal-based and value-based behaviour (Tzafestas, 2016: 25–26). We take the position of Herbert Simon, who says that intelligent machines will over time be able to perform all the functions that are now performed by humans (Tzafestas, 2016: 26). These functions, says Posner (2004), will also relate to creativity, general wisdom and social skills, which, up until just a few years ago, were considered to be exclusively human skills.

How can we tell if a machine is intelligent? The Turing test developed by Alan Turing in 1950 was designed to see if people could differentiate between responses given by a machine and those given by a human (Turing, 1950: 433–460). The test consisted of a set of questions that were asked of a machine and a human, and a human evaluator was to decide whether the answers came from the human or the machine. If the evaluator could not reliably distinguish between the responses from the machine and the human, the test was then said to show that the machine was able to demonstrate intelligent behaviour.

Today, it is generally considered that supercomputers have only been able to surpass humans in two areas: first, this concerned IBM's "Watson", a question-answering computer system, that won the American quiz show, Jeopardy, against human competitors in 2011; second, Deep Blue, an IBM chess-playing computer, that won a chess game in 1996 against the reigning world champion Garry Kasparov (Hsu, 2002). Both of these function areas are related to:

1. The storing and retrieving of large amounts of data, that is, a memory function.
2. Permutations involving large amounts of data, i.e., combinatorics.[6]

On the other hand, no computer has been shown to think strategically and create a vision of a future situation or use intuition in its decision-making; in addition, computers are not able to use the "subconscious" to reach decisions.

Can the intelligent machines we have described above have a moral purpose? If so, are there different moral purposes in the various programmed machines, and do intelligent machines have their own personality? If they do not have their own personality and their own moral purpose, are they anything more than imitating machines, without autonomous decision-making capability? If we say that the first step in understanding morality is an ability to understand the other person's situation and act in solidarity with them – is it then possible that intelligent machines can be ethically aware and act ethically?

Do intelligent robots have an independent moral responsibility? If so, they must first be considered to be independent moral agents. Let us assume that an intelligent robot, which is defined as being just as logical and rationally intelligent as a human, is declared to be a moral agent. This robot is then held responsible for committing an unethical (and illegal) act, and is sentenced to seven years in prison and placed in an isolation cell for robots. This of course sounds absurd – judging, sentencing and imprisoning a robot for committing an unethical (and illegal) act. However, what is the alternative? Who would be held legally responsible, if not the robot? Would it be the designer and programmer, the company that owned the robot, the person who handled the robot, or who else? If the robot is so intelligent that it can reprogram itself on the basis of experience, and by trial and error, then it is unlikely we can punish the programmer, just as one cannot punish the parents for what is done by an adult son or daughter. Just as the parents are responsible for some of the actions and behaviour of their children, but not their adult children, one can imagine that the programmer or company that developed the robot is only responsible for intelligent robots up to a certain level of intelligence and reflection. Beyond this level, the robot is legally considered, metaphorically speaking, to be like an adult son or daughter.

If intelligent robots above a certain level of intelligence are considered to be legally responsible agents, do they then also have rights? If this was the case, then intelligent robots will have gained subject status, that is, they have responsibilities and rights, and they are responsible for their actions. Again, this may sound completely absurd, but, in a not too distant future, questions concerning the above will need to be addressed in relation to singularity. As far as I know, it was Gunkel (2012) who first raised such questions, such as: can machines be ethically responsible? In this book, we choose to formulate the question as follows: under what conditions can intelligent robots have, or not have, an autonomous moral responsibility? The analogy we make here is to parents' responsibility for a child, and how that responsibility falls away when the child becomes an adult.

Both Gunkel's question and our question relate to a historical process where those who have previously been excluded from society with regard to rights, responsibilities, decision-making processes and so on have, at certain points throughout history, been socially included and gained rights and responsibilities. Taking this position, and at the same time linking it to intelligent robots, may be viewed as an innovation in the field of moral philosophy. Moral philosophy has focused on morality and ethics related to humans, and sometimes to animals, but never, as far as I know, to machines. Our contribution to moral philosophy is through reflections on aspects of the answer to the question: under what conditions do intelligent machines have autonomous responsibility? The answers to this question will be discussed in this book.

The philosophical basis for thinking that intelligent machines, under certain conditions, are responsible legal entities and are thus morally responsible for their actions may be related to Agamben's philosophy, especially to his investigations into the legal-philosophical problems regarding "the state of exception" (Agamben, 1998). Agamben says that there are no legal rules that can be adapted to a state of chaos (Agamben, 1998: 16, 18). The point in this context is that all laws and regulations may be viewed as being temporary; they are created in specific situations and changed by new situations. In order for those at the top of the social hierarchy to "bind themselves to the mast", they need a framework to which they can relate; within the national political system, this is often the constitution. In other social hierarchies, there are written and unwritten rules that govern the behaviour of the "socially exalted".

The "exceptions" may be more interesting than the situations that are covered by the rules, says Agamben (1998: 16). It is when exceptional situations occur that the system and the people in the system really show their "character". The saying is that the exception proves the rule. One might also say that a rule's flexibility becomes evident when the exceptions manifest themselves. If you wish to study a rule, a phenomenon or the like, then it is the exceptional situations that should be studied, because this is where we see if the rule is sufficiently flexible and complex. It is by studying the "exception" that various aspects of the "normal" will be highlighted that would not otherwise have been visible. One might say that Agamben's starting point is that the exception has priority in the study of the normal situation.

The good jurist, says Agamben (1998: 16–17), is not the one who has a good memory of the law and is able to apply this to a normal situation, but one who can use their judgement in relation to the intentions of the rules, when the exceptions are to be considered. One can say the same about the good leader – he or she shows their character in situations that become acute, and deviate from the so-called normal.

Regarding legal-philosophical problems, it is the exception that applies when discussing under what conditions intelligent robots will have autonomous responsibility for their actions. The innovative new in the question we have formulated concerns a shifting of perspective in moral philosophy away from the exclusive focus on man. In other words, the anthropocentric perspective changes with the question we have asked above. It is the questions that are the driving force in this book, not necessarily the answers we provide. We attempt to find new knowledge by asking questions related to established knowledge. Our main question in relation to artificial intelligence and ethics is: do we have an answer today to the question of what we really mean by intelligent robots and their moral responsibility? The legal answer lies in the philosophy of the exception. The historical answer lies in the evolution of morality. By the latter, we are referring to the gradual development of autonomous and independent rights throughout history, that is, regarding the position of slaves, serfs, women and other groups who historically did not have autonomous and independent rights, but were considered part of their owners' responsibilities. At various points throughout history, moral perceptions changed, as did relationships regarding rights. The answer to the question above is that we do not have enough knowledge today regarding what will constitute the moral rights of future robots with artificial intelligence. If we go one step further and say that we cannot answer the question before we have an idea of what is meant by moral intelligence, then the question above will be changed to the following: do we today have an answer to the question of what we really mean by intelligent robots and their moral responsibilities related to moral intelligence? Intelligent people without morals can pose a greater problem than intelligent robots without morals. Therefore, moral intelligence is a topic that we will deal with in Chapter 3, where we examine artificial intelligence and ethics.

AI and ethics related to war robots

The question we will examine here is the following: what is the relationship between intelligent war robots and ethics?

"In war, truth is the first casualty", said Aeschylus, 2,500 years ago.[7] Following this approach, ethics is the second casualty of war. The end justifying the means often provides an "ethical" guideline in times of war.

However, we choose here to disregard these negative views on war and instead ask the question: how can military robots, such as drones, utilize ethical intelligence in their design? Research regarding how intelligent systems can equip the military to enable them to gain an advantage over adversaries will always be given huge financial support. Consequently, it is probable that intelligent military robots will be the first to develop super-intelligence, so that they will possess all the necessary available information to enable them to perform multiple functions. These intelligent robots will most probably be connected to other intelligent military systems and have access to continuously updated information.

As mentioned above, ethical considerations are often one of the "casualties" of war. Nevertheless, it may be advantageous for future military intelligent robots to have ethical intelligence built into their design. It is often the case in wartime conflicts that orders get distorted as they move down the command chain, without there necessarily being any evil intent. The orders given by the general staff may become completely distorted by the time they reach the front line, resulting in military actions that are later condemned, something we have witnessed many times, such as in the Vietnam War (Miller, 1978). With the utilization of intelligent military robot systems, the distortion of orders will be taken out of the equation. This will result in the intention that is inherent in the command being given, then being realized in the front line. In this way, the responsibility will remain where it belongs, that is, with the officer or soldier who issued the order. This may result in positive developments regarding the actions of soldiers and officers in the field, since the head of a command chain will need to reflect more on the orders given, because he/she will be aware that they will be held responsible for any actions that are carried out in the field.

There is a current debate about whether or not these intelligent military systems, such as intelligent military robots, should be used in war, because of the highly deadly consequences. Our view is that any weapon that is developed and built will be used in a war conflict at some place and point in time. This applies even if the weapon has been banned for ethical reasons, as we have witnessed with the use of various poisonous chemical agents, cluster bombs and so on. If intelligent military weapon systems are developed and built, they will also be used. Consequently, it is important to discuss the ethical challenges related to the use of such intelligent weapon systems.

From an ethical perspective, all weapon systems have their advantages and disadvantages, such as the example above concerning the distortion of orders moving down through the command chain, as shown in the Vietnam War. While some will welcome a more transparent command chain, where everyone can, in retrospect, see what happened to an order from the top to the front line, there will be others who are critical of such a development. In this context, some will welcome a debate concerning ethics and the use of intelligent weapon systems, while others will be averse to having such a debate, and yet others will be neutral. To illustrate this, we can refer to a simple example – the ethics of the atomic bombing of Nagasaki and Hiroshima by the United States towards the end of the Second World War. From an ethical perspective, many Americans approve the use of atomic bombs, reasoning that it shortened the war and saved many American lives; this viewpoint is also defended by some historians. However, this "American" view contrasts with the view held by many Japanese, who view the bombings as unjustifiable. The point being made here is that in a war there will always be at least two conflicting sides, and different ethical assessments will be made from the positions of the respective parties. This will also be the case with intelligent military weapon systems, once intelligent robots and robot systems become central elements in future military systems.

One ethical defence for developing war robots is that they may deter a potential enemy, preventing the break out of a war. Of course, this is an entirely reasonable argument, because it will save many lives. However, if such weapons are developed and built, they may easily fall into the hands of others than those who developed and used them in the first place. For instance, they may be used by warring states, although it may have been legally established that such weapons must not be sold to warring states. Past experience has shown that warring parties and states will always find someone who is willing to sell them the weapons they want. Therefore, the starting point for an ethical discussion on the development and use of such weapon systems must take into account the fact that they will be used by others for various purposes, in various future conflicts and wars.

The best ethical argument of those who advocate the development of intelligent weapon systems is that they will be able to save the lives of their own soldiers. For instance, consider the pilots of warplanes who risk their lives attacking targets; drone "pilots", on the other hand, will be safe and concealed in bunkers in completely different areas of the world from where the drones are attacking hostile targets. In this context, the geographical distance between the weapon and the operator will also have an impact on ethical considerations, although the obedience principle is an ethical variable in any military context. However, the outcome of the Nuremberg Trials held after the Second World War emphasized the legal and ethical principle that the individual soldier is morally and legally responsible for his/her actions, and cannot hide behind the so-called principle of obedience, i.e., "following orders".

Another argument used in the debate is that the use of intelligent precision weapons, such as war robots, will save lives. Traditional explosive devices will kill anyone who happens to be in the vicinity of the explosion; this may include civil non-combatants such as children, women, the elderly, sick and disabled. On the other hand, the intelligent robots will reduce the injuries and deaths of civilians and damage to buildings, because they will be programmed to take out military targets with precision.

The intelligent war robots will thus be highly lethal weapons, designed to take out any enemy or hostile installation with a high degree of precision. Opponents to their development argue that they should be banned, because their use in war will mean that we are basically moving one step closer to a society where we use our intelligence to kill others.

Yet another point made in the ethical debate concerns the question of responsibility – the question of who will be responsible for any "errors" committed by a robot resulting in the deaths of innocent civilians such as women and children, or in the case where a robot attacks its own forces and civilians. Will it be the designer of the robot who is held responsible? The business that developed and sold the product? The drone pilot? The person who gave the order? We will have difficulties in establishing a clear responsibility for the actions of the robot, and will thus have problems carrying out a judicial process in the event of robots performing illegal actions. If such a robot contravenes the established laws of war, who is then responsible? Can an intelligent robot be held accountable and judged in a court of law? What would the final judgement be? To turn off the electrical power of the robot for ten years? Such a scenario involving the prosecution of a robot in court is not just science fiction, but can obviously lead to the purest nonsense.

A precision surveillance robot, although unable to execute autonomous actions of a lethal nature, can nevertheless contribute to lethal actions, due to the information it provided to a war robot. In the event that the information provided was incorrect, this could result, for instance, in the destruction of a hospital killing innocent civilians. In such a case, would it be possible to subject both the robot and the surveillance robot to judicial procedures, and what would the causality relationship be between the two robots? Who is responsible for what happened? This concerns both legal and ethical problems which, as of today, no one has been able to solve satisfactorily. The military viewpoint is often that some "collateral damage" is acceptable in war conflicts. This may be a commonly held view in the military, but is it acceptable from a legal and ethical viewpoint? Should the military logic determine our ethical understanding, or should our ethical understanding determine the framework of military actions? The outcome of the Nuremberg Trials provides a clear answer to this question. The Nuremberg Trials established that "just following orders" could not be accepted as a defence concerning illegal actions and war crimes. The soldier is responsible for the actions he executes. With such an understanding, we can say that ethical considerations take precedence over military logic. However, even if such a principle were to be established in the case of the use of war robots, we still haven't come any closer to a solution of who is responsible when a war robot makes a serious error that leads to an action that contravenes the rules of war. If we cannot hold the robot responsible, something which we have seen will lead to absurdities, who is then responsible? When we have a problem establishing who is responsible in the case of such illegal actions, the assumption is that such actions will occur more often, because no one is legally responsible. When illegal actions are not subjected to a judicial process, it is not improbable that such abuses will increase. Likewise, one can also envisage an increase in the indiscriminate use of war robots, if no one is held accountable for the illegal actions of war robots. The focus of responsibility is disintegrated in such cases and the actions that lead to the deadly errors will then more easily continue. When responsibility is pulverized in this way, it seems reasonable to assume that the threshold for authorization of illegal actions will be reduced.

Although truth is the first casualty of war, and ethics the second casualty, laws and courts have nevertheless been established over the years to prosecute individuals for war crimes. The Nuremberg Trials after the Second World War was such an institution. The International Criminal Court in The Hague is another institution that has jurisdiction today to prosecute individuals for crimes against humanity and war crimes. When singularity becomes a reality, it is not difficult to envisage ethically intelligent robots assessing the unethical actions of unintelligent human beings.

AI and ethics related to the interaction between people and intelligent robots

What kind of society will emerge when robots take over and replace people in many work functions? Will society become colder and more rational? Will empathy die when robots become ubiquitous in society? Many of these and other similar questions have been asked by scientists and others. The ethical debate regarding the interaction between humans and robots has turned into one of the hottest debates of our time (Nørskov, 2016: 99–100). This debate has arisen both in academia and also among lay people. Perhaps the best known contributor to the debate is the physicist Stephen Hawking.

Almost every debate about intelligent robots is based on an "us and them" dichotomy. When the debate is at its most heated, one could almost wonder if the debate concerned an invasion of Martians rather than the interaction between humans and intelligent robots. It seems that the closer that intelligent robots come to replicating the functions of the human brain, the more the debate takes on an anthropocentric focus. Would the debate have had the same focus if we were talking about an organic hip that was replaced by a sophisticated and intelligent hip prosthesis, or a new technological arm prosthesis that could function exactly the same as an organic limb? Probably not. It seems that when we approach the most important thing about being human, our brain, the ethical debate really takes off. If we manage to produce an artificial heart, an artificial liver, an artificial kidney and so on from a 3D printer, would this lead to the same heated ethical debate regarding a technology that is able to replicate the human brain? What is so special about the brain that it stimulates an emotional debate that envisions apocalyptic scenarios, for example, that the robots will "take over"? Does the debate reach such great heights, because we are actually afraid that we will find out that the human brain is a freak of evolution, a mutation that has led to logical-rational intelligence becoming dominant? If human logical-rational intelligence is a negative mutation, should we fear intelligent robots with a processor capacity of the nth power of the most intelligent human being? Is it perhaps not so much the robots that we fear, but rather ourselves – as logical-rational actors? The idea of "us and them" may be more about ourselves as logical-rational actors. If we remove emotions and empathy from our intelligence, and we are left with logical-rational intelligence, what are we then left with? The mentally ill robot? Is it madness to remove everything except logical rationality and instrumental logic from our thinking? If this is correct, are we not so much afraid of intelligent robots, but rather of ourselves? Do we see our real madness as being mirrored in the logical-rational intelligent robot? Is the ethical debate about intelligent robots really a debate about what we have become, i.e., logical-rational actors in a technological and cold society? Is the fear of intelligent robots just the fear of how we ourselves have evolved as human beings? If this assumption is correct, we can use the ethical discussion about intelligent robots to gain a deeper understanding of what we, as human beings, have now become. Are we logical-rational actors who are only concerned with what can be weighed, measured and counted? The debate about intelligent robots does not concern an "us and them" dichotomy, but is instead a debate about how we see ourselves in the mirror, and how this is mirrored in the intelligent robot. What we see in this mirror is the rational-logical, instrumental actor who only values what can be weighed, measured and counted. We have created the intelligent robot in our own image, and that is what we are afraid of. We are afraid of ourselves. It is perhaps this fear that led to Stephen Hawking, perhaps the most logical-rational intelligent person of our time, expressing sceptical concerns about intelligent robots: in a moment of insight, he might have seen logical-rational people in millions of copies as intelligent robots, carrying out all the functions of private life, work life and society. If logical-rational intelligence becomes the intelligence we value, then it may be said that man has really evolved into a human freak.

The fear of intelligent robots is nothing more than the fear of what we are turning into as humans: rational, logical agents who do not place any value on the intelligence of a smile, emotions or the touch of a hand.

Will our society become a colder one when intelligent robots become commonplace? – this is a question many people ask. The answer is perhaps that we already have a cold society, if we just lift up our eyes and look beyond the church windows and the fences around our own houses. Intelligent robots are not the real ethical challenge, to use Løgstrup's perspective (Løgstrup, 1956); nor is the challenge focused on the "us and them" thinking mentioned above.

The real ethical debate concerns the "ethical mirror". The intelligent robot is the ethical mirror in which we see ourselves mirrored. When we look in this mirror, we see how we have evolved as human beings, and we become afraid. The fear of intelligent robots is the fear of ourselves – the fear that we will remain as we are.

Reflecting on what it means to be human, with our hopes, despair, desires, dreams, expectations, losses, victories and so on, can tell us something about what an intelligent robot can help us with. To only focus on the logical-rational part of our brain is to make a large error when considering the interaction between humans and robots. Rather, a rediscovery of what it means to be a human will be able to add something to the debate about the limitations of the intelligent robot. If being human was only about being logical and rational, we would all have been locked up in mental hospitals a long time ago. Being human is more about the absence of rationality and logical intelligence. To consider people based on technology's development, especially in relation to intelligent robots, does something to the view we have of ourselves. We see ourselves in relation to a very small part of what we are and what we can become. The more we focus on a small part and aspect of being human, the logical-rational, the more we make this part an essential part of being human. When we give logical-rational intelligence primary status, we also give the intelligent robots primary status in our lives. We fear what we see in the ethical mirror – the intelligent robot as a true successor of our elevated logical-rational intelligence. The point here is that if one has chosen to disregard everything except logical-rational intelligence, then one has also turned one's back on being human.

It is not through the type of "machine breaking" that we associate with the Luddites that we can "win" over the intelligent robots, but only by reflecting upon what it means to be human and then understanding how we can use this insight to utilize intelligent robots to help us evolve as humans, and not as ethical mirror images of a logical-rational intelligent robot.

We have illustrated the above discussion in Figure 1.3.

Robots, artificial intelligence, gene–editing and medical ethics

The question we examine here is the following: what ethical issues arise when intelligent robots, artificial intelligence and gene-editing are used in health care?

As we develop new medical technology, we also create new ethical rules, change established rules, adapt to the new rules and cease to use the old rules. Ethical norms and rules are neither imprinted in our genes nor inscribed in stone. They are largely based on practical expediency. When medical robots enter the medical universe, the ethical norms and rules will also adapt to this new development. However, some norms and rules change slowly, while others are created and adapted to technological development. This happens because morality and ethics are largely linked to specific interests and situations.

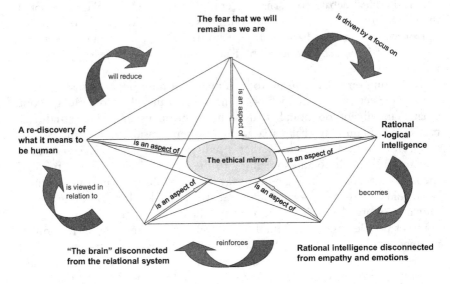

Figure 1.3 The ethical mirror.

The more tightly knit our social systems are, the more adaptable our moral norms and rules will become. Likewise, one can imagine that the more that intelligent robots take over medical functions, the more plastic our medical norms and rules will become. The reason is that the purpose of a medical robot is to diagnose, perform an intervention and follow up the patient. If such a robot can do this better and cheaper than a doctor, then it will make sense to use the medical robot. Consequently, the moral norms and rules that need to be modified to facilitate this change will eventually be established. The purpose of the health sector is to promote, restore and maintain health, and to do this in the most cost-effective way.

There are some clear ethical rules that should be followed in the medical field. These may include:

- Do not create new diseases.
- Do not treat non-existent diseases.
- Do not perform medical experiments that harm people.
- Do not prescribe treatment that has no scientific basis.
- Keep up to date with the latest research on illnesses and diseases.
- Give medical assistance when needed.

Medical expertise is based on recognized scientific principles, which constitute the basis for medical practices in the majority of countries. Shamanism,[8] however, is something completely different from scientific medical practices, and is not included here in the discussion of either medical ethics or the ethical practices that emerge from the interaction between intelligent robots and the medical field. However, we do not say that shamanistic practices cannot have a positive effect on an individual's health – placebos can also improve a person's health. If one has faith in a shaman, this may trigger placebo effects that can have a positive effect on a person's health.

Many physicians take the Hippocratic Oath[9] when they enter medical practice. The Hippocratic Oath is an ethical code that states, amongst other things, that physicians should not harm others physically or mentally, and that they should actively help those who are in physical and mental need.

Medical ethics is changing in step with changes in technology. The technology that broke through during the first industrialization also changed medical ethics. With the emergence of new technology in recent decades, new ethical problems arose. For instance, for how long is it ethically correct to keep people alive using the new technology? Is it a breach of the Hippocratic Oath to disconnect a patient from a respirator? This issue has only increased in complexity with the development of new technology.

New technology has also stimulated the debate regarding physician-assisted death for either terminally ill patients or others, because in reality the tubes and machines of modern medical technology can keep people alive long after they have ceased to function in any meaningful way. The debate on contraception and abortion has also flared up, since new technology makes birth control a simpler practice.

We have chosen to develop ten ethical norms, which indicate ethical challenges related to medical ethics.

Ethical norm 1: Every person has the right to control their own body.
Explanation: You are responsible, if possible, for taking care of your own physical and mental health. It is unethical to exploit other people either physically or mentally. This applies both to corporal punishment and other forms of physical violence against people, and also to the death penalty. Murder is always murder, although it may be sanctioned by the law. Ethical norm 1 deals with rights and obligations in relation to one's own health.

Ethical norm 2: Every person, without exception, is entitled to medical care.
Explanation: For those who are unable to obtain health insurance, the government has an obligation to take care of their medical treatment, without it costing the individual anything. Cooperation between private and public health providers can be an advantage in many cases, but must not result in those who cannot afford health care having to pay.

Ethical norm 3: Everyone is responsible for taking care of their own health.
Explanation: Those who do not take care of their own health directly and indirectly harm other people. This can be explained by the fact that they may become large consumers of hospital services, thus burdening health service resources and contributing to hospital queues becoming longer. In this way, they inflict unnecessary suffering on others. This may apply to those who are obese, abuse alcohol, are inactive and others.

Ethical norm 4: Everyone has an obligation to help others promote their health.
Explanation: Helping others to promote their health also helps to reduce the pressure on the healthcare services. Thus, more people will be able to get help faster as and when they need it.

Ethical norm 5: Everyone is responsible for ensuring a health-promoting environment.
Explanation: Those who pollute the environment also injure the health of others directly or indirectly, and in the short or long term,

Ethical norm 6: Healthcare personnel are obliged to exclusively use treatments that are scientifically supported.
Explanation: This norm is included to prevent quacks and frauds from penetrating the health system. The purpose of the norm is to ensure good health for the entire population, and to prevent quackery and other non-scientific medical practices from entering the health services. One approach to prevent or limit quacks and frauds from entering the health services is through:

a.   Legislation, and
b.   ensuring that the state offers equal medical assistance to everyone who is resident in the country.

We know from research that placebo treatment helps in roughly 30 per cent of cases.[10] However, is it ethical to prescribe a placebo? It may be scientifically proven that a placebo can work in some cases, but is it medically and ethically justifiable to use this knowledge to help a patient? If one thinks that one should never lie to a patient, then prescribing a placebo will be wrong. If, on the other hand, one believes that white lies can be useful to help a patient, then it may be ethically defendable to prescribe placebos.

Ethical norm 7: Healthcare professionals are obliged to strive for optimal patient safety and gain the patient's authorization to undergo any medical intervention, wherever possible.
Explanation: The patient has the right to be engaged in decisions regarding their treatment, especially when their life is at risk, and should in most cases be heard. Very ill or injured persons may have difficulty in exercising this right. In this case, their fate will lie in the hands of health personnel and/or close family. However, there will be special circumstances in relation to this ethical norm such as in disasters of various kinds, where decisions will have to be made as to who will be given priority regarding treatment. For instance, will the young be given priority in front of the elderly? Should one give priority to those who are most seriously injured? Intuitively, most people would probably give priority to those who have the greatest chance of survival, regardless of age.

Ethical norm 8: Health personnel have a duty to protect patients from interference by fanatical political movements and extreme religious sects.

Explanation: Medical personnel who participate in experiments on prisoners, for example in prison camps, not only violate medical ethics, but may also be involved in crimes against humanity. For instance, such experiments were performed in Japanese and German concentration camps during the Second World War. We also have reports of similar occurrences in Argentine and Chilean prisons during the periods when the military junta ruled in those countries.

There are also reports of unethical and harmful medical experiments being performed by the CIA amongst others, where those subjected to the experiments were injured for life. Furthermore, there are also examples of pharmaceutical companies having conducted unethical experiments on people, with harmful results (Bunge, 2013: 200). All of these examples breach the ethical norm that medical personnel have a duty to protect patients.

Ethical norm 9: Health and medical workers have the right to join professional organizations.

Explanation: This ethical norm is in accordance with the International Labour Organization's (ILO) Declaration on Fundamental Principles and Rights at Work.

Ethical norm 10: Health workers have the right to protect themselves against unfair accusations of medical malpractices.

Explanation: Such a norm safeguards the rights of healthcare professionals and, through its consequences, makes medical treatment safer.

Conclusion

We have examined the following question in this chapter: how can we understand, explain and apply ideas about ethics related to intelligent robots and artificial intelligence?

If one wishes to design ethically reflecting robots, one should perhaps start by considering how ethically reflecting people would behave and act in concrete situations and cases. This sounds plausible, but is it correct? Could it rather be the case that intelligent robots of the future will have a greater capacity for ethical reflection than humans? An individual will often put personal motives, gain and self-interest before what he/she thinks is the ethically right thing to do in a situation. By contrast, an intelligent robot can be developed, designed and programmed so that ethically correct actions take precedence in any context and situation. In this way, the answer to the above question is yes – in the future, intelligent robots will have a greater capacity for ethical reflection than humans.

We must be able to trust that an intelligent robot does what is considered ethically correct in every situation, even though we cannot be sure that people would do the same. However, there is a paradox here. Can we rely on the robot to carry out ethical actions when the person using the robot is an ethically challenged person? In such a situation, can it be conceived that personal ethics overrides the ethics of the intelligent robot? If so, should the intelligent robots then be programmed so that they cannot be overridden by the individual's personal ethics? If we do the latter, we will also open up the possibility that intelligent robots take over the decision-making in the choice of actions, at least in relation to ethical issues.

Is it the case that the ethical standards programmed into intelligent robots should be accepted by most people, or only by those who own the robot, or those using the robot? Should there be a particular ethical perspective and some preferred ethical theories included in the programming? The questions sound complex and difficult to solve. In this context, we can ask: what would the individual human being do in a specific context and situation? The question and answer lead us to a simpler way of looking at ethics and morals, that is, it is within the specific context and situation that ethics will be applied. The same could be expected of the intelligent robot. Understood in this way, the design principles are simple, even although it will be time consuming to programme the robot. Based on the specific contexts in which the intelligent robot will operate, it will be necessary to think through and implement action rules for all conceivable situations that may arise. Ultimately, there will always be situations that have not been thought through, but which the robot will face. In such situations, one should use a feedback system that integrates learning into the robot's programmed algorithms, so that the intelligent robot can learn from experience.

One can further assume that robot ethics will be linked to the goals, tasks and functions that the specific robot will perform. For instance, the ethics that apply to a self-driving car will not necessarily be the same as those that apply to a war drone in a military operation. With this in mind, robot ethics will be relative and linked to context, goals and functions of the different intelligent robots, as will also be the case for people in different situations.

As we have seen from the discussion in this chapter, it seems reasonable to assume that we need a framework for how people and robots interact and act ethically in various contexts and situations.

The answer to the question we have examined in this chapter can be hierarchically divided into three levels:

Level 1 concerns context and culture. This level is designed for the more general ethics (artificial general intelligence). At this level, norms and values are programmed into the overall program of the robot. This is the level of meta-ethics where questions are asked concerning what makes ethical truths true.

Level 2 is the situational level, and concerns applied ethics, where ethics is adapted to the individual situation. For instance, we are concerned with how ethical theories and principles are applied in practice. The ethical basis at this level concerns relationships and understanding of patterns.

Level 3 is the operational level, where rules and procedures are applicable within the specific relevant context and situation. This is the level at which normative ethics operates. Normative ethics is concerned with what one should or should not do in concrete situations.

We have shown these three levels in Figure 1.4.

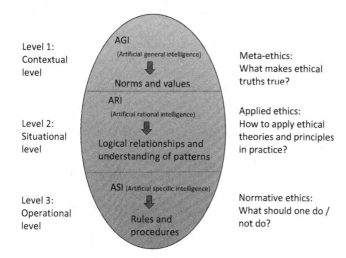

Figure 1.4 Framework: the hierarchical levels of ethics for intelligent robots.

Notes

1 Here we use the word "informat" to refer to intelligent robots that are linked to other intelligent robots by means of a global network. This network allows the robots to be continually updated.
2 RRD stands for respect, responsibility, dignity.
3 https://en.wikipedia.org/wiki/Technological_singularity.
4 https://en.wikipedia.org/wiki/Singularity_Summit.
5 The saying originates from https://en.wikipedia.org/wiki/The_road_to_hell_is_paved_with_good_intentions. Bernard of Clairvaux is known as the founder of the Cistercian Order and wrote the outlines of the Rule of the Knights Templar, https://en.wikipedia.org/wiki/Bernard_of_Clairvaux.
6 https://en.wikipedia.org/wiki/Combinatorics.
7 https://en.wikipedia.org/wiki/Aeschylus.
8 https://en.wikipedia.org/wiki/Shamanism.
9 https://en.wikipedia.org/wiki/Hippocratic_Oath.
10 https://en.wikipedia.org/wiki/Placebo.

Bibiography

Agamben, G. (1998). Homo sacer, Stanford University Press, Stanford.
Allen, C.; Wallach, W. & Smit, I. (2006). Why machine ethics? IEEE Intelligent Systems, 21, 4: 12–17.
Asimov, I. (2008). I, robot, Bantam, New York.
Benhabib, S. (2004). The rights of others: Aliens, residents, and citizens, Cambridge University Press, Cambridge.
Bunge, M. (2013). Medical philosophy, World Scientific, London.

Cave, P. (2002). Responsibility in law and morality. Hart Publishing, New York.

Coeckelbergh, M. (2010). Health care, capabilities, and AI assistive technologies, Ethical Theory and Moral Practice, 13, 2: 181–190.

Coeckelbergh, M. (2014). The moral standing of machines: Towards a relational and non-Cartesian moral hermeneutics, Philosophy and Technology, 27, 1: 61–77.

De Grey, A. (2013). The curate's egg of anti-aging bioethics, in More, M. & Vita-More N., (eds.). The transhumanist reader, Wiley-Blackwell, New York. pp. 214–219.

Doudna, J. & Sternberg, A. (2018). A crack in creation: The new power to control evolution, Vintage, New York.

Dreyfus, H.L. (1979). What computers can do: The limits of artificial intelligence, Harper Colophon Books, New York.

Gert, B. (1988). Morality, Oxford University Press, Oxford.

Gips, J. (1992). Towards the ethical robot, in Ford, K.; Glymour, C. & Mayes, P. (eds.). Android epistemology, MIT Press, Cambridge. pp. 243–252.

Gunkel, D.J. (2012). The machine question: Critical perspectives on AI, Robots, and Ethics. The MIT Press, Cambridge, MA.

Gunkel, D.J. (2014). A vindication of the rights of machines, Philosophy and Technology, 27, 1: 113–132.

Hall, J. (2000). Ethics for machines, In: Anderson, M.; Leigh, M. & Anderson, S. (eds.). Machine ethics, Cambridge University Press, Cambridge. pp. 28–46.

Hsu, F.H. (2002). Behind deep blue: Building the computer that defeated the world chess champion, Princeton University Press, Princeton.

Løgstrup, K.E. (1956). Den etiske fordring, Gyldendal, Copenhagen.

McCarthy, J. & Hayes, P.J. (1969). Some philosophical problems from the standpoint of artificial intelligence, in Meltzer, B. & Michie, B. (eds.). Machine intelligence, 4th edn. Edinburgh University Press, Edinburgh. pp. 463–502.

Miller, J.G. (1978). Living systems, McGraw-Hill, New York.

Nørskov, M. (2016). Technological dangers and the potential of human-robot interaction: A philosophical investigation of fundamental epistemological mechanisms of discrimination, in Nørskov, M. (ed.). Social robots, Ashgate, London. pp. 99–121.

Operto, V.G. (2011). Roboethics: A bottom up interdisciplinary discourse in the field of applied ethics in robotics, IRIE, 6, 12: 2–8.

Posner, R. (2004). Catastrophe risk and response, Oxford University Press, Oxford.

Rodogno, R. (2016). Robots and the limits of morality, in Nørskov, M. Social robots: Boundaries, potential, challenges, Ashgate, New York. pp. 39–57.

Ronson, J. (2011). The psychopath test, Picador, New York.

Shanahan, M. (2015). The technological singularity, MIT Press, Boston.

Shaun, N. & Knobe, J. (2007). Moral responsibility and determinism: The cognitive science of folk intuitions, Noûs. 41, 4: 663–685.

Turing, A. (1950). Computing machinery and intelligence, MIND, 49: 433–460.

Tzafestas, S.G. (2016). An introduction to robophilosophy, River Publishers, New York.

Veruggio, G. (2005). The birth of roboethics, in Proceedings of IEEE international conference of robotics and automation (ICRA 2005), Barcelona. pp. 1–4.

Wallach, W. & Allen, C. (2009). Moral machines: Teaching robots right from wrong, Oxford University Press, Oxford.

2   Robots and ethics

The key ideas of the chapter

1. Intelligent robots can have morals.
2. In certain circumstances, intelligent robots can take autonomous decisions.
3. Under certain conditions, intelligent robots have moral responsibility.
4. All organizations that develop artificial intelligence and intelligent robots should be subject to a legal requirement to establish an ethics board. Such a board would review the research being conducted in those areas where the organization develops and produces products. Statutory ethics boards should operate at the same level as the qualified auditors that audit an organization's financial affairs.
5. A robot has moral responsibility for its actions when it has a high level of moral reflectivity and is designed to learn from its own actions.

Introduction

Driverless cars, tractors, trains, planes, drones and so on are only the first step towards the development of intelligent robots.[1] Only once we experience the reality of artificial intelligence, artificial emotions and artificial relationships will we truly understand the scope of intelligent robots, which will go on to become intelligent informats.[2]

The development of revolutionary new technology tends to cause a kind of ripple effect in society. Initially this effect will be limited to the small area of society where the technology is adopted. Many years may pass before the new technology becomes dominant in society. This happened with the steam engine, the combustion engine, railways, electricity, electronics,[3] computers, etc. When the new technology finally becomes generally accepted, many changes will already have occurred in the old methods of production, distribution patterns and consumer behaviour.

It is highly likely that some of these new innovations, including, for example, singularity technology and singularity innovation (Johannessen, 2020), will bring about economic and social crises. People who lose their jobs, or who are left with skills that are no longer required as a result of the new technology, will be hit hard (Johannessen, 2018a; 2018b). Schumpeter called this creative destruction (Johannessen, 2017): the creation of the new results in the destruction of the old.

The purpose of this chapter is to examine the ethical consequences of such a development, i.e., how intelligent robots will take over many of the functions previously performed by humans. The most high-profile cases where ethical problems have materialized involve driverless cars (Mitchell & Wilson, 2016). Can a driverless car be legally liable? Can it be prosecuted if it injures a human being? Who should pay? One could imagine that the situation would be the same as where a person driving a car injures a third party. The driver will be insured, and damages to compensate for the third party's injuries will be paid by driver's insurance company. Just as having no insurance is not an option when buying a car, for example, so a court could find a robot, as a moral agent, legally liable, since ultimately it would be the insurance company that would pay the damages. Today it may seem far-fetched to think of a court holding an intelligent robot, e.g., a Tesla driverless car, liable, but perhaps it is no more unlikely than a human being found guilty of unlawful driving (Lin, 2016).

The current ethical debate about driverless cars will no doubt set the standard for many of the intelligent robots that will take over many different functions in the years to come. Accordingly, driverless cars make a good case study when considering ethical problems associated with intelligent robots (Lin et al., 2017: ix).

If driverless cars, despite the negative examples we have all heard about, can save more lives on the roads, shouldn't they be developed and implemented on a large scale? The same applies to intelligent medical robots. If these robots can save more lives than a human surgeon, shouldn't they be developed for full-scale application at hospitals, even though they may make mistakes? If using military robots, e.g., drones, can save more lives than not using them, shouldn't they be put into operation? The immediate objection is that not all ethical issues can be weighed, measured and counted in the number of human lives saved. Human rights can be difficult to quantify, for example. Although beta-versions of technologies may make mistakes, we have to remember that humans make mistakes too. This applies to drivers on the roads, surgeons in operating theatres, army officers in wartime and so on. When using a beta-version of a technology, one must anticipate that the technology will be imperfect.

It is not the case that one can follow the law and assume that one is acting ethically. Throughout history, many laws have turned out to be an ethical disgrace when viewed with hindsight. Slavery is just one example of a situation that was legal, but that subsequently was declared immoral. Child labour is another. There are also laws and regulations that one has to break in certain situations if one is to act ethically. Jumping a red light or crossing a solid double line on the road may be the correct decision if it means getting a critically ill child to hospital sooner, and thereby increasing the chances of saving the child's life. Law and ethics are related, but they are not necessarily correlated or causally related. We must be cautious both when technology is changing and about allowing the law to take precedence over ethics. Particularly in the case of new technology, for example, technological singularity (Johannessen, 2020), a legal basis does not yet exist. In such cases, one must resort to the legal philosophy of the exception and the limitations of the law (Agamben, 1998; 2005; 2013a; 2013b) on the one hand, and ethical reflections on the other hand.

Science and technology[4] are an ethical project (Bunge, 1985: 58). According to Herbert Simon, the Nobel Prize-winner and pioneer in artificial intelligence, anyone who is involved in science has a responsibility: "to assess, and try to inform others of the possible social consequences of the research products he is trying to create" (Simon, 1991: 274). One could take that a step further, and say that all organizations that develop artificial intelligence and intelligent robots should be subject to a legal requirement to establish an ethics board. Such a board would review the research being conducted in the areas where the organization develops and produces products (Burell, 2014). Ethics boards are nothing new. They have existed in the pharmaceutical sector since the 1980s (Clark, 2006). Organizations that develop intelligent robots and artificial intelligence could learn from the experiences of these existing ethics boards. The purpose of such ethics boards would be to reflect on the implications of the design and use of intelligent robots. Think about the following questions: should intelligent robots tell white lies? If an intelligent car has a choice between running over a child or an elderly person, what choice should it make? Should a military drone kill innocent civilians, the elderly, women and children, if that means a terrorist leader will also be killed? It would be neither possible nor appropriate for individual engineers or programmers to provide clear answers to such questions. There are many pressing and critical questions that an ethics board could consider in order to find theoretical, conceptual and practical answers. As Sullins (2016: 90) also suggests, an ethics board should engage in a dialogue with designers, programmers and possibly end-users. In that way one would be able to have more confidence that the reflections of the ethics board were not based simply on a superficial ethical assessment.

It is possible that it is too early to roll out ethics boards in all the organizations involved in the development of artificial intelligence. Nevertheless, it is our experience that it is better to experiment with ethics boards before a technology goes past a point that makes it too late to discuss what could have been done to prevent an undesirable development. Our point is thus that statutory ethics boards should operate at the same level as the qualified auditors that audit an organization's financial affairs. Such audits allow an organization's financial course to be corrected if necessary, at an early stage. Statutory ethics boards could also operate in this kind of way.

There is a long history of technical tools taking over tasks that are laborious for humans. In the Iliad, Homer describes Hephaestus, the Greek god of technology, who creates "golden servants" for humans (Lattimore, 1961). More than 2,000 years later, Leonardo da Vinci described a mechanical object similar to what we would today call a robot (Hill, 1984). Today, science fiction is full of

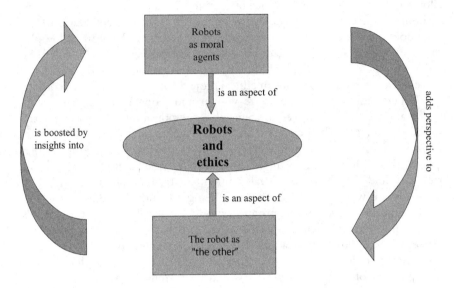

Figure 2.1 Aspects of robot ethics.

descriptions of robots taking over human functions and replace humans on the planet (Wilson, 2005).

The point of this brief historical summary of the origins and development of robots is to see them in the perspective of what a human is, which we don't find in any well-known stories about robots, either in Homer or sci-fi literature. This is human beings, with their moral brains and moral intelligence. If nothing else distinguishes humans from intelligent robots, then it is this: humans' moral intelligence.

The question we are investigating here is as follows: what is robot ethics? We looked briefly at this question in Chapter 1. In this chapter, we investigate this question more thoroughly.

To answer this question, we have developed two sub-questions:

1. How do robots as moral agents relate to robot ethics?
2. How do robots as the "Other" relate to robot ethics?

In Figure 2.1, we have summarized the introduction. Figure 2.1 also shows how this chapter is structured.

Robots as moral agents

The question we are investigating in this section is as follows: How do robots as moral agents relate to robot ethics? To find the answer to this question, we have chosen to answer four sub-questions:

1. What is a moral agent?
2. Subject to what criteria should intelligent robots make their own moral decisions?
3. Can and should intelligent robots be independently liable for their decisions?
4. Do intelligent autonomous robots have rights?

What is a moral agent?

In general, moral situations involve a moral agent. But, what exactly is a moral agent? In principle, we can say that a moral action consists of two parts. The person or thing who executes the moral action, and the person or thing who is at the receiving end of the action. In this relationship, the moral agent is the party who executes the action. Both parties in this moral situation have rights and responsibilities. If an error occurs in the execution of the action, the party affected can demand compensation in one way or another. But is it reasonable to compensate a machine, even if it is intelligent? If an intelligent robot has a right, then it seems reasonable that the robot can receive compensation. This hypothesis links two concepts: rights and compensation. What we are saying here is that if one has a right, then one is also potentially entitled to compensation. If this statement is given meaning and interpreted sensibly, then we are facing an emergent concept in moral philosophy: the intelligent robot as moral agent. If an intelligent robot can be understood as a moral agent, then we should also reflect on whether and why human beings have a greater value than an autonomous intelligent robot.

If an intelligent robot is considered to be a moral agent, will it then also be responsible for its actions? Today, such a question may seem absurd, but when technological singularity occurs, it will become a relevant question. One of the implications of the question is that those programming these intelligent robots will need to reflect on what are the right and wrong actions that such an intelligent robot may have to choose between in various situations. For example, both autonomous and integrated intelligent robots may be part-systems in a car. In a situation where an accident seems unavoidable, the car may have to make a decision between colliding with a pregnant woman or a small five-year-old child. How should the car's safety systems reach a decision in such a case? To complicate the example, we can say that the woman is the driver's wife, while the child is a stranger. These and similar questions are rapidly leaving the realm of science fiction and becoming part of everyday reality. However, before situations such as this become a commonplace reality, we need to first reflect on the composition of the program that will be installed in intelligent robots.

Is it only humans who can possess morality? From a Kantian philosophical perspective, the answer to this question is yes (Gunkel, 2012: 3). In 1971, three of Professor Peter Singer's students from Oxford University published a book (Regan, 1999: xi) that focuses on animal liberation (Singer, 2015). Animal liberation concerns, amongst other things, the moral status of animals – can animals be moral agents? From our standpoint, this may be understood as a turning point, where thought is given to whether or not non-humans, in this case animals, can be moral agents – the next logical step would be to consider if intelligent robots can also be moral agents. Of course, determining whether animals and intelligent robots can be moral agents, and thus be responsible for their actions, is dependent upon how we define morality. According to the Kantian definition, animals and intelligent robots are not moral agents, because in the Kantian view morality is: "the rational determination of the will" (Gunkel, 2012: 3). Animals are thus, by this definition, excluded. However, we will here use the philosopher Mario Bunge's definition of morality, which is the art of "helping others enjoy life" (Bunge, 2010: 4). With such a definition of morality, both animals and intelligent robots can possess morality, and thus be understood to be moral agents. One consequence of such an understanding is that we as humans have moral obligations towards both animals and intelligent robots. The innovative new, however, is that both animals and intelligent machines also have a moral commitment to humans (Hall, 2007). As we have shown above, situations can easily arise where an intelligent robot has to make decisions involving moral considerations, such as telling white lies, or in the situation where a car has to choose between two people, when a collision is unavoidable.

Using Bunge's definition of morality, can a dog and an intelligent robot "help others enjoy life"? Anyone who has owned a pet will most probably answer yes to this question, and anyone who has had an intelligent prosthetic limb fitted, such as an arm or a leg, will also give a positive answer to the question. For the answer to the question of what a moral agent is – this is here related to Bunge's definition of morality, except that the moral agent is acting as a subject or object. However, we do not know specifically what intelligent robots will be capable of in the future. Nevertheless, many of them may be able to "help others enjoy life". Intelligent robots can therefore be called moral agents, even though today this may seem absurd and like science fiction. Although we do not fully know what intelligent robots will be capable of in the future, we know that they will have the ability to be moral agents, given the definition of morality we use here. To distinguish human morality from intelligent robots as moral agents, one can define the morality of intelligent robots as being artificial morality. This make a functional distinction between humans and intelligent robots in relation to morality. However, we do not assume that human morality is superior to the artificial morality of the future intelligent robot. On the contrary, in many cases, artificial morality will be superior to human morality. In this context, it seems unnecessary to refer to the numerous examples of poor human moral behaviour throughout history.

Under what conditions should intelligent robots make their own moral decisions?

If intelligent robots using artificial intelligence can be defined as moral agents, will they also be able to make their own autonomous decisions? Before we can answer a clear yes to this question, there are some conditions that must be present. These conditions concern the degree of logical-rational intelligence and the degree of emotional intelligence.

In some cases, an intelligent robot will have a high degree of both logical-rational intelligence and emotional intelligence. In such cases, the intelligent robot has the same moral status as a human being, and can therefore make autonomous moral decisions. We have discussed such intelligent robots in the book *Singularity Innovation* (Johannessen, 2020).

At other times, an intelligent robot will have a high degree of logical-rational intelligence, but a low degree of emotional intelligence. Under such conditions, the intelligent robot cannot make moral decisions, because one cannot expect the robot to have sufficient capability both to understand and empathize with others. In such situations, one can say that the intelligent robot can make semi-autonomous decisions. In practice, this means that a person with a high degree of emotional intelligence should control the moral decisions of the robot, and should help it to consider the emotional aspects of these decisions.

In cases where the robot does not have a high degree of logical-rational intelligence or emotional intelligence, it will not be able to make any autonomous decisions. In such a case, one can hardly speak of intelligent robots at all, but more of a type of "vacuum cleaner robot". Such robots have pre-programmed tasks that they perform and they can hardly be described as intelligent. We have included them here because they are still robots, and are the type of robot that is most common today.

Some robots may have a high degree of emotional intelligence, but a low degree of logical-rational intelligence, a combination that can also be found in humans. It is therefore also important to determine the moral status of such robots. In situations where these robots need to make decisions, we believe that they can only make decisions under the control of someone with a relatively high degree of logical-rational intelligence. We also define these as semi-autonomous decisions.

We have illustrated the above description with a typology in Figure 2.2.

Can and should intelligent robots have autonomous responsibility regarding their decision-making?

Can humans disclaim responsibility in the case of intelligent robots that can make autonomous decisions, as shown in Figure 2.2? In the other three cells of Figure 2.2, it seems reasonable to say that people cannot disclaim their responsibility. The point that is not clear is the question of responsibility concerning the highly intelligent robots that also have a high emotional intelligence. From the 1990s until the present, a new branch of research has been developing to answer this type of question, which goes under the name of Artificial Moral Agents (AMA) (Gips, 1991; Wallach & Allen, 2009; Allen & Wallach, 2014; White & Baum, 2017: 67–79).

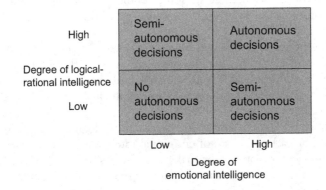

Figure 2.2 Moral decisions by intelligent robots.

Today, most intelligent robots are ethically blind, i.e., they do not have ethical intelligence programmed into their decision-making capacity. However, this will change when robots are introduced that can make autonomous decisions, because these will have built-in emotional intelligence (Johannessen, 2020). Our point of departure is that if intelligent robots make autonomous decisions, they will then also have responsibility for their actions as moral agents. Two examples that clearly relate to this statement are military robots in the air, water and on land, as well as self-driving cars. However, we will add here that although intelligent robots make autonomous decisions and are responsible for their actions, any claims for damages are not necessarily directly related to the decision made and the allocation of responsibility, because here, as elsewhere, this will be an insurance matter. If a self-driving car makes a moral error, or a correct moral assessment, which nevertheless has fatal consequences for some but not for others, then one can argue that the car is responsible, but that the compensation will be paid by an insurance company. It will then become a legal question as to whether it is the car owner, the car manufacturer, the designer, etc., against which the insurance company seeks possible redress. The point in our case is that the car as a legal object is responsible for its autonomous decisions.

Why should we design and build intelligent robots that are able to make moral decisions and that have responsibility for their actions? The answer is simple: we already use robots with artificial intelligence that are able to make autonomous decisions such as self-driving cars that may have to choose who should or should not be injured if a collision is unavoidable. Although we have already built and use intelligent robots, we lag behind on the development of the emotional and moral aspects of intelligent robots. If an intelligent robot that can make autonomous decisions only has logical-rational intelligence built into its design without ethical intelligence, it should immediately be admitted to the nearest "mental asylum for robots", because such a robot is very dangerous. It is only when the emotional and moral component is built into the design of the robot that it can be declared well and fit and allowed back into society to continue its logical operations.

It is highly improbable that the development and construction of intelligent robots will be discontinued at any point in the future, despite the fact that their use may result in serious injuries to people. Nonetheless, we still have the opportunity today to reflect on how to develop robots with a capacity to make moral decisions, so that in the future they can have integrated artificial emotional intelligence.

It can be imagined that an intelligent robot capable of making autonomous decisions will have different levels of moral reflection (cf. Figure 2.2). It can also be imagined that the autonomous decisions of the intelligent robot operate on two logical levels. On the first level, the intelligent robot uses the moral code of the designer and/or user. On the second level, the robot is disconnected from the designer's and user's moral code, but learns through trial and error the nature of the functional morality related to the various contexts and situations. We have illustrated this in Figure 2.3.

If a robot can be held responsible for an action, then we face some very difficult legal problems, including the distribution of responsibility. For instance, there is the question of the distribution of responsibility between the robot, its owner, the designer and the company that designed the robot, or parts of it. The distribution of responsibility, however, is a legal problem, and, for all practical purposes, an insurance problem that will be solved by the insurance companies and their lawyers. Our point in this section is clearly shown in Figure 2.3: the robot is morally responsible for its actions when it has a high degree of moral reflection and when it is designed to learn from its own experience. Compensation in a particular case will be a legal, not an ethical, problem.

Not all robots are responsible for their actions, as shown in Figure 2.3. This is not surprising, because it is also the case that not all people are responsible

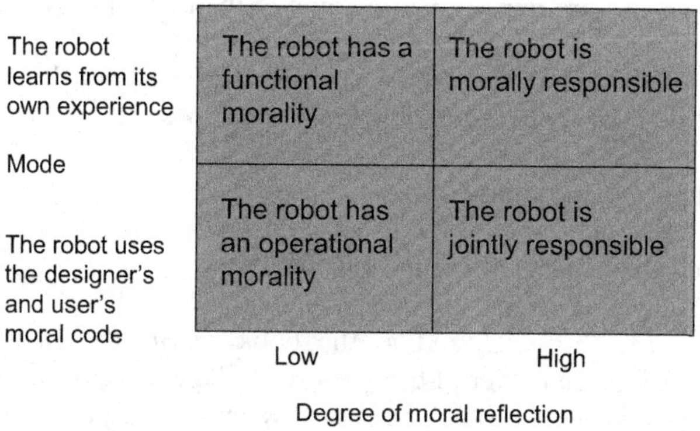

Mode	Low	High
The robot learns from its own experience	The robot has a functional morality	The robot is morally responsible
The robot uses the designer's and user's moral code	The robot has an operational morality	The robot is jointly responsible

Degree of moral reflection

Figure 2.3 Autonomous decisions by an intelligent robot and the robot's moral responsibility.

for their actions. Children and those who are of unsound mind are not held responsible for their actions and are treated differently from others.

Do intelligent autonomous robots have rights?

Does a hammer have morals? No, of course not. The hammer has no characteristics that make it comparable to a human being. The above question is not as ridiculous as it sounds on first reflection. All legislation in the West is based on the fact that robots do not have obligations or rights. Robots are not legal persons and are therefore not part of the legal system (Tzafestas, 2016: 75). However, we are not discussing the legal aspects of robots here, but the moral and ethical aspects.

A person who dies as a result of a robot hugging the person too hard will have been involved in an accident according to today's laws. However, with the emergence of intelligent robots with some form of self-consciousness, empathy and emotional intelligence, will this be a morally correct legal assessment? The turning point for whether robots will have rights like humans is related to whether they can develop consciousness (Singer & Sagan, 2009). We have discussed emotional and relational robots in the book *Singularity Innovation* (Johannessen, 2020). It seems reasonable to assume that intelligent robots, with emotional and relational intelligence, should have the same rights as animals.

Can animals possess morality and be responsible for their actions? Hubbard (2011: 405–441) discusses exactly this problem in his article: "Do androids dream? Personal and intelligent artefacts". Amongst other things, Hubbard says that animals are responsible for their actions within certain conditions. Of course, it is the owner who is financially responsible for the animals he owns; in this connection, there are many countries that legally require dogs to be insured, so any accidents will result in an insurance case. There is clearly an analogy between intelligent robots that are morally responsible for their actions and, for example, dogs that are responsible for some of their actions (White & Baum. 2017: 70).

Hubbard (2011: 405–441) highlights three elements that make non-humans, i.e., animals and intelligent robots, responsible for their actions:

1. They have complex relationship skills.
2. They are self-conscious.
3. They understand the benefits of group affiliation.

As we have seen in Chapter 1 and in the book *Singularity Innovation* (Johannessen, 2020), it is just a matter of time before intelligent robots develop artificial intelligent relational and emotional skills. When this happens, some time after singularity has occurred in 2040 (Kurzweil, 2005; 2008; Levesque, 2017), an intelligent robot will meet Hubbard's requirements that will establish that robots are responsible for their actions. If we assume then that under certain conditions robots will be responsible for their actions, does this imply that they will also have rights?

Peter Singer, one of the most influential moral philosophers of our time, addresses this question in his book *Ethics in the Real World* (2016: 291–294). He starts by describing a situation where a fully automated robot helps an elderly woman to manage in the home, so she will not have to be admitted to a nursing home. The elderly woman is clearly very happy with the help that the robot can give her. We can also imagine that such a future intelligent robot will help the elderly woman in every way possible to fill her needs for food, social relations, sex, etc. (Levy, 2009). Such a robot will then fulfil Hubbard's (2011) three requirements for being responsible for its actions. If one is responsible for one's actions, is it not natural that one also has rights? Robots already make cars and help the elderly so they can carry on living at home. Service robots also perform some of the cleaning jobs in the home and other places. The logical, social and emotional intelligence of robots is evolving and increasing very quickly. In the near future, we will have humanoid robots in our homes that can help us with our various tasks and needs. Robots that can replace babysitters are not far off in the future. Such robots will soon also be able to express empathy and possess a very high level of relational skills (Johannessen, 2020; Lin et al., 2014). Therefore, these robots will not only possess logical intelligence. It is these robots to which we are referring when we ask the question of whether robots should have rights, not to robots of the vacuum cleaner type or at the level of a mechanical hammer.

What we do need to be aware of, at least at a moral level, is that robots that can automate many functions in the home and the workplace will eventually force many unskilled and skilled workers into unemployment. Will the intelligent robot have any moral responsibility for this future situation? Should the robot have a right to compete with unskilled workers? These questions may sound meaningless today, but they will soon become more relevant in the near future. Should the robots have rights, so that a new "Luddite" movement does not crush them with a moral hammer, or worse, a leaden hammer of fury?

Will future humanoid sex robots be able to help marriages survive, or will it lead to more divorces? Will singularity cause humans to become completely inferior to robots?

Should the robots in such a situation have legal rights? The potential for loss of control of working life, in a situation where intelligent robots take over most job functions, is certainly a realistic future scenario. If such a scenario were to occur, would it be the robots or people who should be guaranteed fundamental rights? It doesn't take much imagination to visualize humans feeling threatened by the intelligent robots we have described above. It is then highly probable that various types of Luddite movements will emerge. Some of these movements may be based on the fear of loss of work, while others may be based on the fear of moral degeneration being brought about by "love and sex robots". Whatever the basis of these Luddite movements, it seems reasonable to assume that the robots may be subjected to an attempt to comprehensively

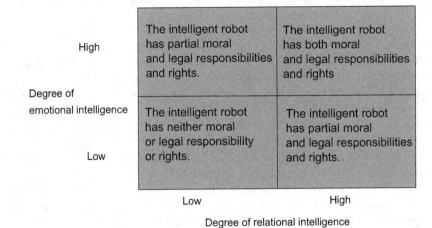

Figure 2.4 Under what conditions should intelligent robots have rights?

destroy them. Should robots be given moral and legal rights before such a situation occurs in the future? It took a long time before animals were given rights and protected by the law (Singer, 2015). Should intelligent robots have the same rights as a puppy, a kitten, etc.?

Given the above discussion, it seems reasonable to now reword our original question to: under what conditions should intelligent robots have rights? The tentative answer is illustrated in Figure 2.4.

The robot as the "Other"

In this section, we will focus on a particular understanding of intelligent robots – specifically, in terms of what we will term the "Other". This is a term taken from the French philosopher Simon de Beauvoir's book *The Second Sex*.

In our presentation of an intelligent robot as the "Other", there are five conditions that must be met:

1. First, in order for an intelligent robot to be termed the "Other", it must be able to pass a modern version of the Turing test (Gunkel, 2012: 55–65).
2. Second, it must be able to demonstrate some ability in forming relationships (Gunkel, 2012: 93–157).
3. Third, it must be able to differentiate between social inclusion and social exclusion (Gunkel, 2012: 159–221).
4. Fourth, it must possess certain "personal" characteristics (Cheok et al., 2017: 193–213).
5. Fifth, the intelligent robot must be able to satisfy a human sexually (Levy, 2009).
6. Sixth, it must possess moral intelligence.

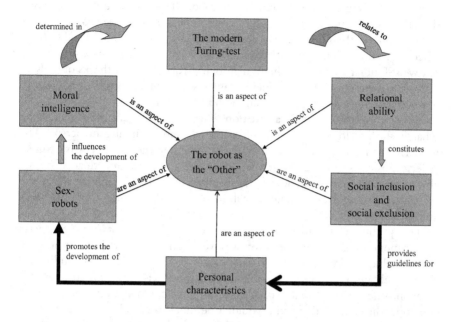

Figure 2.5 Aspects of an intelligent robot as the "Other".

We have shown figuratively in Figure 2.5 what we believe the conditions are that must be met in order for an intelligent robot to be termed the "Other". Figure 2.5 also shows how we have organized this section.

The modern Turing test

The Turing test, developed by Alan Turing in 1950, was designed to see if people could differentiate between responses given by a machine and those given by a human. The Turing test is a behavioural test: i.e., if a machine behaves intelligently, then it is intelligent.[5]

How can one know if a robot, or for that matter another person, thinks and feels emotions? Churchland had already asked this question 20 years ago (1999: 67). One can take this a step further by asking: how can we know that another person possesses consciousness? The point we are making here is to question the full validity of the classic Turing test. For instance, perhaps a modern version of the Turing test should also include querying how one can find out if a human being thinks and possesses consciousness. If it is not possible to determine this how can, then how we know if an intelligent robot thinks and feels? Although scanning programs have been developed in neuroscience that can identify certain processes occurring inside the brain when a person thinks and feels (Nunez, 2016), we still don't fully understand how the human brain works, or what the "mind" is. This obviously poses problems in developing a modern Turing test for an intelligent robot. If we do not know whether the other person with whom we are communicating possesses consciousness, thinks and feels, then how can we develop a modern Turing test related to these aspects for an intelligent robot? We may know how a computer system works, but we still don't know how the human brain works when it thinks or feels, or what consciousness is. Despite these problems, we will nevertheless attempt to develop a modern Turing test for an intelligent robot.

We can risk ending up in a situation where an intelligent robot reflects on what it should do when it is confronted with an unintelligent person. The modern Turing test will aim to find out whether the other person thinks, reflects, possesses moral intelligence and has the ability to empathize. If we are able to develop such a test, developing a similar test for an intelligent robot will be relatively straightforward, i.e., it will involve the design and development of an elementary algorithm.

The point we are making here is that we should not impose stricter requirements on an intelligent robot than we do on a human being. For humans, it is possible that we will never get beyond having faith that an individual thinks, feels and is conscious at some level. It is not really possible to get inside someone's head to see what's going on. Even if we could, figuratively, see inside someone's head, then there still wouldn't be much we could find out, because we wouldn't know what to look for. When we do not know how we think, how emotion is produced and what consciousness is, what would we then be looking for inside the other person's head? We have taken a roundabout route back to the original Turing test. That is, we just have to accept that what we can get to know within the foreseeable future is linked to behaviour – which the original Turing test revealed. We can assume and imagine that the other person thinks and feels, but we can only know this from his/her behaviour. Why would it be different for an intelligent robot? Our only opportunity to assess the intelligent robot, as for a human, is to enter into communication and then observe the reaction.

We know that the software of an intelligent robot manipulates symbols and imitates actions, events and people. But, how can we know that this is not exactly the same as what a human being does? When we do not know how people think and feel, how can we say that a person thinks and feels differently from an intelligent robot? Now that we have reflected on aspects of similarities and possible differences between a human and an intelligent robot, we can ask the next question: how can we uncover the moral level of an intelligent robot, or a human being? We will assume here that the answer to the question is linked to relational ability as well as to the capacity to socially include and exclude others. In the next two sections, we will discuss these points.

Relational capacity

Does an intelligent robot have relational capacity, that is, the ability to develop, maintain and deepen relationships? Although some people may have problems with social interaction, such as those suffering from Asperger syndrome, an autism spectrum disorder,[6] most people are able to form social relationships. Some may be better at this than others, but as a general rule, most people are able to develop and maintain relationships. We have already referred to the fact that robots used in Japanese elderly care have some form of relational capacity. It might be argued here that it is not so much the robots that have relational capacity, but rather that they trigger a relational competence in the patient. On the other hand, if we envisage that an intelligent robot in the future will soon have developed emotional and relationship skills (Johannessen, 2020), then the intelligent robot will also have relational capacity. If we further disregard science fiction literature and imagine that in the future it will be possible to develop a material that mimics human skin, then the intelligent robot will have taken a new step towards developing relational capacity by imitating human characteristics.

If we look at relational capacity from the robot's perspective, can it be conceived that the robot will have a moral responsibility towards the human it is supposed to serve? If the intelligent robot has a moral responsibility towards a human, can it also possess consciousness? Should the intelligent robot be developed so that its relational capacity includes the three elements of respect, responsibility and dignity? Benhabib says that these three elements are necessary, if not sufficient, in order to exhibit moral behaviour (2004; 2011). If we are able to develop intelligent robots that have relational capacity and a built-in moral standard, where the three above-mentioned elements are an integral part of the design, then we will have taken a step away from the ideas of the literature that considers intelligent robots in terms of the "Frankenstein Complex" (Asimov, 1983: 62; 1985; 2008).

When the roles of the intelligent robot, and the person who has designed it, are in doubt, we will of course have a problem. No one knows if this is possible. However, we do know with great certainty that if this were to happen, then a Frankenstein monster will indeed have been created by humans. To avoid this development, we should already now be requiring that those who develop artificial intelligence, and artificial emotional and relational intelligence, should integrate the three elements of respect, responsibility and dignity into all intelligent robot systems. If we do not incorporate some form of moral intelligence into our intelligent robots, then we will most certainty build a future "Frankenstein system". If we only develop logical-rational intelligence, we may exponentially increase the productivity of the robots and their functions in the workplace, but if this is increased without emotional and moral intelligence, we will simultaneously be creating a future Frankenstein system. Once this has occurred, it will be too late to change anything. Therefore, now is the time to take on the role of a modern Prometheus, the Greek God who was able to look into the future, because it will be too late when we can only look back to "the point of no return", when it once was possible to change something.

Hubris is often said to be the downfall of humanity. However, it may be too late this time, if we do not use moral intelligence to strengthen the logical-rational intelligence of the robots. Our view is that it is not only immoral in itself to omit moral intelligence in the design of intelligent robots but also immoral because this will amount to leaving the intelligent robots to their own devices in a technological environment, without any moral map and compass. The robots will not be responsible for us – it is rather us who are responsible for them. Therefore, we should not leave them to a cold world, but should instead integrate moral intelligence into their design to the same degree as we integrate logical-rational intelligence into their design. In future scenarios, an intelligent robot that only possesses logical-rational intelligence will eventually be dismantled, because if the robot had been a human being, such a person would have been taken away by two strong men in white coats to a suitable institution.

Unfortunately, it is not only human hubris that will create such a Frankenstein monster, that is, an intelligent robot without emotional, relational and moral intelligence. We also know that many people lack moral intelligence. This is where the real "Frankenstein complex" is buried. Consequently, we should worry about people's lack of moral intelligence, because morally unintelligent people are hardly likely to worry about intelligent robots being built without moral intelligence. In their eagerness to design logical-rational intelligent robots, people have forgotten that they also need to reflect on their own moral intelligence. The first step on the road to moral intelligence is to understand the importance of, and conditions for, social inclusion and exclusion. Throughout humanity's history, social inclusion has been intimately associated with well-being and personal growth, while inappropriate social exclusion can have disastrous effects. For instance, the Greek philosopher Socrates, accused of corrupting Athenian youth, was given the choice of death by drinking poison from a cup, or exile to the Greek colonies in the Black Sea. He chose to drink the cup of poison rather than be socially excluded from his friends and family in Athens.

Social inclusion and exclusion

It seems reasonable to assume that the interaction between humans and intelligent robots will become qualitatively different in the near future, from the relationship between technology and humans is today. The robots will increasingly imitate human behaviour in many ways, and most will probably gain an understanding of the phenomenon of social inclusion and exclusion. Imagine the following scenario: you invite an intelligent humanoid robot to dance at a social event, but you are declined. Would this rejection be experienced in the same way as if one had been rejected by a person? Would it be morally correct of the robot to reject an invitation to dance? Is it morally correct of an intelligent robot to reject a human being in any situation? If we consider a robot as a social partner to a human being, should we expect the robot to be loyal to the partner? Results from several researchers have found that when the robot's appearance and behaviour are human-like, the robot will be accepted on the same level as if it were human (Kanda et al., 2007: 962–971; Fong et al., 2003: 143–166). Research has also found that robots are socially included when they exhibit a diversity of human-like social behaviour (Breazeal, 2004: 182).

Of course, although an intelligent robot possesses logical, emotional and relational intelligence, it is still not an actual human being. The point is that the robot will behave in human-like ways. Not least, it is important to understand that anyone who needs a social robot for whatever reason, social, sexual, communicative and so on, will be able to relate to the robot as if it were a human being. Of course, one objection is that ideally humans should primarily seek contact with other humans. However, the fact is that there are hundreds of millions of humans in the world who experience loneliness.[7] In addition, there is also gender imbalance in various parts of the world, that is, a disparity between males and females in a population. The reasons for this are varied, such as war conflicts and the one-child policy in China, to mention but two. This means that there is a large unmet need for social robots that can fulfil some of the needs of people who are alone for various reasons. What we need, says Brezeal (2004: 182), are social partners, even though they may be intelligent robots.

If the robots evolve with logical, emotional, social and moral intelligence, will they represent a new type of living system? Will we have a moral obligation towards this emergent system? Will these robots be social "individuals" that are able to include and exclude people? Will humans be able to include and exclude the robots? We know that the above description may resemble a science fiction movie; however, in a few years' time, singularity will occur. In the period following singularity, it is probable that moral, social and emotional intelligence will be integrated into the robots that already possess logical-rational intelligence. It is in relation to a most probable future scenario that we ask these questions here. This type of robot, which we term an Agora robot in the book *Singularity Innovation* (Johannessen, 2020), will be able to understand people, although most people will not fully understand how this robot is designed and self-taught, and is able to modify its own programming. When an Agora robot understands people, can it then also socially include and exclude people? We have developed a typology that responds to aspects of this question in Figure 2.6.

Without robots being able to understand people and vice versa, no social inclusion is possible. On the other hand, it is quite possible to socially exclude, when we do not understand the "Other". To put it more clearly, it is precisely when we do not understand the other that the other is viewed as "a stranger", and we have a tendency to socially exclude strangers. We need to understand: the diversity of signals that the other sends out, the context and the signals the context sends out and finally the situation and the signals the situation sends out. If this basic communicative process can be interpreted by a future Agora robot, then it seems reasonable to assume that such a robot will have its own personal characteristics that will allow it to include and exclude people on the basis of certain selection criteria.

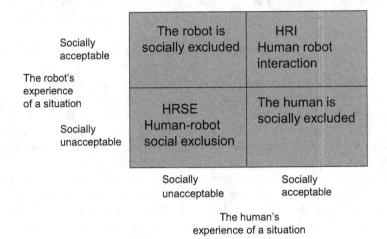

Figure 2.6 Robots and social inclusion and exclusion.

Personal characteristics

Can a robot have personal characteristics? On the basis of these characteristics, can a robot include and exclude people. Can a robot, based on its personal characteristics, like some people better than others? If a robot has personal characteristics, can it then be called a person? A person is of great importance to us, because it is a construct built up from moral values. But what are the characteristics of the constructed person? The question is: what are the characteristics that constitute a person? If we are able to find the necessary and sufficient characteristics that constitute a person, then we will need to find out if an intelligent robot can also possess these characteristics.

Many of those who try to answer the question of what constitutes a person mention two characteristics: rationality and self-consciousness (Singer, 1999: 87; Kadlac, 2009: 422). Others, such as Smith (2010: 54), list as many as 30 characteristics that constitute a person. We will simplify the characteristics that constitute a person, and make use of necessary and sufficient characteristics, in line with Dennett's (1998: 268) schematization. As already mentioned, the necessary characteristics are rationality and self-consciousness. However, these are not enough. There are also the sufficient characteristics, which are: emotional, relational and moral intelligence. We have illustrated this in Figure 2.7.

Out of all the five conditions in Figure 2.7, self-consciousness may be the most difficult for an intelligent robot to develop. Self-consciousness is perhaps the most important human characteristic (Locke, 1996: 146; Himma, 2009: 19). If the robot can develop all other characteristics, but not self-consciousness, is it possible to characterize the robot as a person? However,

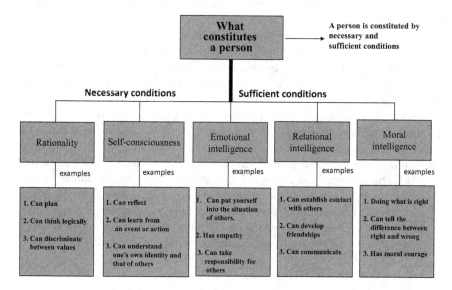

Figure 2.7 What constitutes a person.

viewing future intelligent robots as mindless automatons, machines that can only perform pre-programmed actions, will be wrong. A future intelligent robot will be rational, and have developed aspects of emotional, relational and moral intelligence. The question is, however, whether it will be able to develop self-consciousness. In order to approach this question, we must first know more about human consciousness. At present, what constitutes human consciousness is still an open philosophical question (Bedau, 2014; Bedau et al., 2008; Blok et al., 1997).

Sex robots

Is it morally wrong to have an intimate relationship with an intelligent robot? Is it morally wrong if a romantic relationship develops between a human and an intelligent robot?

The first international scientific conference on love and sex with robots was held in Madeira in 2014 (Cheok, et al., 2017: 193).

Levy (2007a) mentions five characteristics that define romantic and intimate relationships:

1. Intimacy
2. Frequent interdependence
3. Personal connection
4. Affection derived from similarity between the partners
5. Sexual attraction

It should be possible in the future to design an intelligent robot that is able to exhibit the five characteristics listed above in relation to a human partner. New technology is being developed that can simulate "touch and feeling communication" (Cheok, et al., 2017: 196–198), which may be viewed as a prerequisite for future robots, if they are to be able to demonstrate Levy's five characteristics.

Levy (2007a) claims that the crucial point in relation to whether we should treat the robots from an ethical-moral perspective is whether they can be said to be conscious. Therefore, according to Levy, the degree of consciousness is a prerequisite, regarding whether we should adopt an ethical perspective in our relationship with future robots. We support Levy's view here, but with the reservation that we will consider the degree of consciousness. If the robots fulfil the necessary and sufficient conditions in Figure 2.7, they should be treated like human beings. Though this may sound strange today, the robots of the future that possess the characteristics of a human will be viewed by people from an ethical perspective.

It was Levy who first discussed the idea of robots as sex objects and prostitutes. He suggested the following ethical challenges related to sex robots (2007b):

1. The ethics of using robots as sex objects.
2. The ethics related to the individual who uses a sex robot.
3. The ethics concerning the individual and society regarding the use of robot prostitutes.
4. The ethics of using sex robots as prostitutes.
5. The ethics concerning one's partner if one uses robot prostitutes.
6. The ethics regarding human sex workers when people use robot prostitutes (i.e., in relation to taking a part of their market).
7. The ethics related to the actual sex robots.

It is also possible to view sex robots from a different perspective. These robots will probably be programmed to manipulate the emotions of the human beings who will use them as sex machines. From this perspective, one can argue that the sex robots will go beyond what would be ethically acceptable to a human being, since the robot is programmed to manipulate a person's emotional life. From this perspective, Sullins (2012: 398–409) claims that there should be restrictions on the robot industry in relation to how far they can go in manipulating the emotional lives of people.

Are you cheating on your partner if you have sex with a robot prostitute? If the answer is yes, what are the limits of fidelity and infidelity? For instance, is masturbating included? Or using sex toys, such as a dildo? As Levy has pointed out on several occasions, the answer is a clear no; one is not unfaithful if one uses sex robots to satisfy oneself, just as someone is not unfaithful if they use a sex toy such as a dildo.

Moral intelligence

We will define moral intelligence here as the system of expressed values, values in action and empathy. This definition takes us from knowing what is right and wrong to acting in the morally right way. It is a practical definition that both states which elements constitute moral intelligence, while also emphasizing that moral skills and empathy are action-oriented.

Our values are based on our assumptions, and it is presumed that they guide our behaviour. However, empirical research has shown that in reality there is only a tenuous link between our values and our actual behaviour (Hechter et al., 1993; Boyatzis et al., 2000: 47–64). Therefore, moral skills and empathy are important elements when showing that moral intelligence is related to our "operating philosophy" (Goleman et al., 2002). Our values in operation, when carrying out actions, are related to our moral skills (Argyris & Schon, 1982). Empathy, the ability to understand the feelings of others, is an essential part of moral intelligence (Goleman, 1995; 1998).

Moral intelligence is based on fundamental and universal values. This means that moral intelligence, as we use the concept here, is not linked to cultural relativism. Cultural relativism is the view that an individual's values are relative to the culture to which they belong (Taylor, 1991). We assume here that there are some fundamental moral values that are common to all people such as respect, responsibility, dignity, perceived justice and decency (Benhabib, 2002; 2004; Benhabib et al., 2006).

Some researchers claim that morality and intelligence are two different domains that cannot be linked (Bellaby, 2014: 1–5). However, we are of the opinion that intelligence without morality and morality without intelligence are on par with an airplane without wings or a boat without a rudder. Moral or ethical intelligence is like optimism, something that can be learned and developed, a view shared by Dobrin (2002: 38). We know from countless historical examples that people with a high intelligence quotient (IQ) do not necessarily develop a high moral intelligence. We also know from several historical examples that those with a high moral intelligence do not necessarily have a high IQ.

In this way, the concept of moral intelligence appears as a paradox.[8] The paradox, however, finds its solution in an inseparable connection between ethics and intelligence. This does not mean that you need to be intelligent to act morally, or act morally because you have a high IQ. It means that the connection tries to show that moral intelligence is a separate form of intelligence, which emerges in the connection between the two elements.

How can you know what high moral intelligence is? One can learn much from the answer Aristotle gave to the question of what is a good person. Aristotle asked the questioner to find a righteous person and then observe what this person does (Dobrin, 2002: 37). This method connects moral intelligence to practice, reflection and action.

One takes as a starting point the values that apply to all humanity such as respect, responsibility, dignity, perceived justice and decency (Benhabib, 2002; 2004; Benhabib et al., 2006). One can then assess how the individual uses these values in practice. Finally, the individual's empathy plays an important role, because this says something about how the universal values are incorporated into the individual's actions. Understood in this way, the behaviour of the person in practice will determine their moral intelligence.

This approach is closely related to Lawrence Kohlberg's theory of moral development (Garz, 2009). Kohlberg's theory is based on the close relationship between moral reasoning and moral actions (Gibbs, 2013; Kohlberg, 1981; 1984).

We base our own thinking and ideas concerning universal values on Benhabib's ideas and philosophy regarding multicultural values (Benhabib, 2002; 2004; Benhabib et al., 2006).

In the global knowledge economy, there is much to suggest that we should utilize multicultural or universal values, because our cohesive strength is found precisely in our fundamental values, even though everything else separates us.

Our identity and belonging are essentially different, but our values can be the cohesive strength that binds us together despite differences in language, sexual orientation, ethnicity, nationality, gender, "race" and so on (Benhabib, 2002: 24–49). In our view, our common values are universal and represent a cohesive strength, for which Benhabib et al. (2006) argue. We share these universal values with all of humanity, not just with those who share our nationality. Such universal values stress the importance of treating everyone with the same **dignity**, no matter which position and situation the person is in. A second value that is universal is that no matter when and where we meet a person, everyone should be treated with the same **respect** (Benhabib, 2002: 11–108). The third universal value we will emphasize here is **responsibility** (Benhabib, 2004: 104). By this, we mean that wherever the other person(s) is, we as moral individuals have a responsibility for his/her life situation.

It is this thinking regarding the concept of moral intelligence that we use when viewing the intelligent robot as the "Other".

Conclusion

We have investigated the following questions in this chapter: what is robot ethics?

Robot ethics seems to be developing as a new research area at the interface of robot design and ethics. This new research area investigates the ethics of everything, from robots in the entertainment industry, to robots in the sex industry, in military operations, in the operating theatre, in self-driving cars, self-driving trains and buses, self-driving planes and so on.

In this chapter, we have looked at the conditions that must exist for a robot to be able to take autonomous decisions. We have also investigated whether a robot can be morally responsible for its actions or, more precisely: in what circumstances can one say that a robot is responsible for its actions? In this context, we also examined whether a robot has legal rights.

If we develop intelligent robots that have logical/rational intelligence, emotional intelligence and relational intelligence, does that mean that in certain situations the robot will have both moral and legal responsibility? We also investigated this question in this chapter. We developed the concept of moral intelligence in order to give a more detailed answer to the general question: what is robot ethics? By reflecting on the concept of moral intelligence, we can also reflect more closely on the question of the degree of morality in individual human beings. In other words, it is not only robots that must develop moral intelligence, humans must do so too.

Robot ethics is an area within the field of applied ethics, i.e., the study of the relationship between ethics and technology. In addition, robot ethics comprises a system of ethics, technology, science and society. In this way, the new scientific field of robot ethics can derive elements of knowledge from psychology, jurisprudence, sociology and other areas.

The question posed by this chapter can best be answered by describing robot ethics as "a work in progress". However, in this chapter we have added some important elements to this "work in progress". This is particularly true in respect of the following areas of robot ethics:

1. Under what conditions can robots take autonomous decisions?
2. Under what conditions does a robot have moral responsibility?
3. Under what conditions does a robot have moral and legal rights?
4. Under what conditions can a robot be viewed as the "Other"?
5. Under what conditions can we view an intelligent robot as "a person"?

Why is the new discipline of robot ethics important? The main reason is that technology will soon make it possible to design robots based on molecular science, which means that these robots will be terrifyingly similar to humans. It may take 10, 20, 30 or 40 years before we reach this stage of singularity, but technological developments are pointing definitively in this direction. Accordingly, we must ensure that it is not simply the moral codes of the robot designers and software developers that are implemented into these intelligent robots. Ethics must be based on a level of knowledge that is thoroughly considered and reflected over, and where moral intelligence is an important factor. Accordingly, we cannot leave robot ethics to software developers, just as we cannot leave war to the generals.[9]

Notes

1 The word robot was first used in 1921 in a play by the Czech author Karel Čapek. The robots in the play were humans made of flesh and blood, by lacking any creativity or feelings. Robot means "serf labour" in Czech (https://en.wikipedia.org/wiki/Karel_Čapek).
2 An intelligent informat is an intelligent robot that is linked to, and continually updated via, a global information network.
3 https://no.wikipedia.org/wiki/Elektronikk.
4 Technology is defined here as: "The scientific study of artefacts" (Bunge, 1985: 219).
5 https://en.wikipedia.org/wiki/Turing_test.
6 https://en.wikipedia.org/wiki/Asperger_syndrome.
7 To take a few disparate examples: In Denmark alone, in 2015, 1.6 million single people were reported to be living alone (www.dst.dk/en/Statistik/nyt/NytHtml?cid=19041).

18 per cent of all adults live alone in Norway (http://ledernett.no/nesten-1-av-5-boralalene). 50.1 per cent live alone on Manhattan (New York (https://nypost.com/2009/10/30/poll-half-of-manhattan-residents-live-alone/). Of the 49,062,530 households in Japan, 14,457,083 households are one-person households, and the percentage of people living alone is 29.47% (https://stats-japan.com/t/kiji/11902).
8 Quinlan (2007: 1) refers to Claridge, who says that intelligence ethics is an oxymoron, i.e., a figure of speech in which contradictory terms appear together.
9 This is based on a remark by the French statesman Charles-Maurice de Talleyrand.

Bibliography

Agamben, G. (1998). Homo sacer, Stanford University Press, Stanford.
Agamben, G. (2005). State of exception, Stanford University Press, Stanford.
Agamben, G. (2013a). The highest poverty, Stanford University Press, Stanford.
Agamben, G. (2013b). Opus Dei: An archaeology of duty, Stanford University Press, Stanford.
Allen, C. & Wallach, W. (2014). Moral machines: Contradiction in terms or abdication of human responsibility? In Lin, P.; Abney, K. & Bekey, G.A. (eds.). Robot ethics, MIT Press, Cambridge. pp. 55–68.
Argyris, C. & Schon, D. (1982). Theory in practice learning, Jossey-Bass, San Francisco.
Asimov, I. (1983). Asimov on science fiction, HarperCollins, New York
Asimov, I. (1985). Robots and empire, Doubleday, New York.
Asimov, I. (2008). I, Robot, Bantam Books, New York.
Bedau, M.A. (2014). Artificial life, in Frankish, K. & Ramsey, W. (eds.). The Cambridge handbook of artificial intelligence, Cambridge University Press, Cambridge, pp. 296–315.
Bedau, M.A. & Humphreys, P. (2008). Emergence: Contemporary readings in philosophy and science, MIT Press, Boston, MA.
Bellaby, R.W. (2014). The ethics of intelligence, Routledge, London.
Benhabib, S. (2002). The claims of culture, Princeton University Press, Princeton.
Benhabib, S. (2004). The rights of others: Aliens, residents, and citizens, Cambridge University Press, Cambridge.
Benhabib, S. (2011). Dignity in adversity: Human rights in troubled times, Polity Press, New York.
Benhabib, S.; Waldron, J.; Honig, B. & Kymlicka, W. (eds.) (2006). Another cosmopolitanism, Oxford University Press, Oxford.
Blok, N.; Flanagan, O. & Güzeldere, G. (1997). The nature of consciousness, MIT Press, Cambridge, MA.
Boyatzis, R.E.; Murphy, A.J. & Wheeler, J.V. (2000). Philosophy as a missing link between values and behavior, Psychological Reports, 86: 47–64.
Breazeal, C. (2004). Social interaction in HRI: The robot view, IEEE Transaction on Systems, Man and Cybernetics Part C Applications and Reviews, 34, 2: 191–196.
Bunge, M. (1985). Philosophy of science and technology. Part I: Epistemology & methodology III, Dordrecht: Reidel.
Bunge, M. (2010). Political philosophy, Dordrecht: Reidel.
Burrell, I. (2014). Google buys UK artificial intelligence startup Deepmind for £400m, The Independent, January 27.

Cheok, A.D.; Karunanayaka, K. & Zhang, Y.E.Y. (2017). Lovotics: Human-robot love and sex relationships, in Lin, P.; Jenkins, R. & Abney, K. (eds.). Robot ethics 2.0, Oxford University Press, Oxford. pp. 194–213.

Churchland, P.M. (1999). Matter and consciousness, MIT Press, Cambridge, MA.

Clark, H. (2006). Chief ethics officers: Who needs them? Forbes, October 23.

Dennett, D.C. (1998). Brainstorms: Philosophical essays on mind and psychology, MIT Press, Cambridge, MA.

Dobrin, A. (2002). Ethics for everyone: How to increase your moral intelligence, John Wiley & Sons, New York.

Fong, T.; Nourbakhsh, I. & Dautenhahn, K. (2003). A survey of socially interactive robots, Robotics and Autonomous Systems, 42, 3/4: 143–166.

Garz, D. (2009). Lawrence Kohlberg: An introduction, Verlag Barbara Budrich, Berlin.

Gibbs, J.C. (2013). Moral development and reality: Beyond the theories of Kohlberg, Hoffman and Haidt, Oxford University Press, Oxford.

Gips, J. (1991). Towards the ethical robot. In Ford, K.G.; Glymour, C. & Hayes, P.J. (eds.). Android Epistemology, MIT Press, Cambridge. pp. 243–252.

Goleman, D. (1995). Emotional intelligence: Why it can matter more than IQ, Bantam, New York.

Goleman, D. (1998). Working with emotional intelligence, Bantam, New York.

Goleman, D.; Boyatzis, R. & McKee, A. (2002). Primal leadership: Realizing the power of emotional intelligence, Harvard Business School Press, Boston.

Gunkel, D.J. (2012). The machine question: Critical perspectives on AI, Robots, and Ethics. The MIT Press, Cambridge, MA.

Hall, J.S. (2007). Beyond AI: Creating the consciousness of the machine, Prometheus Books, Amherst.

Hechter, M.; Nadel, L. & Michod, R.E. (1993). The origin of values, Aldine de Gruyter, New York.

Hill, D. (1984). A history of engineering in medieval and classical times, Croom Helm, London.

Himma, K.E. (2009). Artificial agency, consciousness, and the criteria for moral agency: What properties must an artificial agent have to be a moral agent? Ethics and Information Technology, 11, 1: 19–29.

Hubbard, F.P. (2011). Do androids dream? Personal and intelligent artefacts, Temple Law Review, 83: 233–243.

Johannessen, J-A. (2017). Innovation leads to economic crises, Palgrave, London.

Johannessen, J-A. (2018a). Automation, innovation and economic crisis: Surviving the fourth industrial revolution (Routledge studies in the economics of innovation), Routledge, London.

Johannessen, J-A. (2018b). The workplace of the future: The fourth industrial revolution, the precariat and the death of hierarchies (Routledge studies in the economics of innovation), Routledge, London.

Johannessen, J-A. (2020). Singularity innovation (Routledge studies in the economics of innovation), Routledge, London.

Kadlac, A. (2009). Humanizing personhood, Ethical Theory and Moral Practice, 13, 4: 421–437.

Kanda, T.; Sato, R.; Saiwaki, N. & Ishuguro, H. (2007). A two month field trial in an elementary school for long term human robot interaction, IEEE Transactions on Robotics, 23, 5: 962–971.

Kohlberg, L. (1981). The Philosophy of moral development: Moral stages and the idea of justice, Vol. 1, Harper Collins, New York.

Kohlberg, L. (1984). Psychology of moral development, Vol. 2, Joanna Cotler Books, New York.

Kurzweil, R. (2005). The singularity is near, Penguin, London.

Kurzweil, R. (2008). The age of spiritual machines: When computers exceed human intelligence, Penguin, London.

Lattimore, R. (trans.). (1961). The Iliad of Homer, University of Chicago Press, Chicago, IL.

Levesque, H.J. (2017). Common sense, the Turing test, and the quest for real AI, The MIT Press, Boston, MA.

Levy, D. (2007a). Love and sex with robots: The evolution of human-robot relationships, HarperCollins, New York.

Levy, D. (2007b). Robot prostitutes as alternatives to human sex workers, Proceedings of the IEEE-RAS International Conference on Robotics and Automation (ICRA 2007), April 10–14, Rome. (www.prostitutionresearch.com/wp-content/uploads/2012/01/Robot-Prostitutes-as-Alternatives.pdf)

Levy, D. (2009). Love and sex with robots, Gerald Duckworth, New York.

Lin, P. (2016). Is Tesla responsible for the deadly crash on auto-pilot? Maybe, Forbes, July 1.

Lin, P.; Abney, K. & Bekey, G.A. (2014). Robot ethics: The ethical and social implications of robotics, The MIT Press, Boston, MA.

Lin, P.; Jenkins, R. & Abney, K. (2017). Robot ethics 2.0, Oxford University Press, Oxford.

Locke, J. (1996). An essay concerning human understanding, Hackett, Indianapolis, IN.

Mitchell, R. & Wilson, S. (2016). Beyond Uber, Volvo and Ford: Other automakers plan for self-driving vehicles, Los Angeles Times, August 19.

Nunez, P.L. (2016). The new science of consciousness, Prometheus, New York.

Quinlan, M. (2007). Just intelligence: Prolegomena to an ethical theory, Intelligence and National Security, Vol. 22, 1–13.

Regan, T. (1999). Foreword. In Steve, P. (Ed.) Animal others: On ethics, ontology, and animal life. SUNY Press, New York. pp. xi–xiii.

Simon, H.A. (1991). Models of my life, Basic Books, New York.

Singer, P. (1999). Practical ethics, Cambridge University Press, Cambridge.

Singer, P. (2015). Animal liberation, Bodley Head, New York.

Singer, P. (2016). Ethics in the real world, Princeton University Press, Princeton.

Singer, P. & Segan, P. (2009). Do humanoid robots deserve to have rights? The Japan Times, 17 December 2009.

Smith, C. (2010). What is a person? Rethinking Humanity, social life and the moral good from the person up, University of Chicago Press, Chicago.

Sullins, J.P. (2012). Robots, love, and sex: The ethics of building a love machine, IEEE Transactions on Affective Computing, 3, 4: 398–409.

Sullins, J.P. (2016). Ethics Boards for research in robotics and artificial intelligence: Is it too soon to act? In Nørskov, M. (Ed.). Social robot, Ashgate, London. pp. 83–98.

Taylor, C. (1991). The ethics of authenticity, Harvard University Press, Boston.

Tzafestas, S.G. (2016). Roboethics: A navigating overview, Springer, Heidelberg.

Wallach, W. & Allen, C. (2009). Moral machines: Teaching robots right from wrong, Oxford University Press, Oxford.

White, T.N. & Baum, S.D. (2017). Liability for present and future robotics technology, in Lin, P.; Jenkins, R. & Abney, K. (eds.). Robot ethics 2.0, Oxford University Press, Oxford. pp. 66–79.

Wilson, D.H. (2005). How to survive a robot uprising: Tips on defending yourself against the coming rebellion, Bloomsbury, New York.

3 AI and robot ethics

The key ideas of this chapter

1. Reflection on synthetic ethics.
2. The theoretical point here is that ethical reflection should take place in advance of the development of artificial intelligence. If ethical reflection takes place in the wake of an action done by a robot, it is very likely that the ethical perspective will lose out to arguments about efficiency.

Introduction

The idea that artificial intelligence and intelligent robots will come to dominate humans has led both scientific and science fiction authors to speculate as to whether we will reach to a point in our development where intelligent machines will start to think about what they should do with unintelligent humans. Such a development is not only frightening, but apocalyptic. We may think it imaginary, speculative and unrealistic.

However, it is extremely realistic to think that the artificial intelligence in intelligent robots could be abused by people who are more motivated by their own interests than ethical reflection. Regardless of whether this or some other scenario actually materializes, it is very important for us to reflect on artificial intelligence and ethics.

The artificial intelligence that will be developed in the near future will take the form of small "nanorobots". These will be designed at molecular level and will be invisible to most of us. These "nanorobots" will probably be designed like Lego bricks, so that they can be adapted and applied in various kinds of super-robots, with various functions and purposes (Boden, 2016; Franklin, 2017).

Nanorobots will operate in most of our everyday activities, from opening and closing doors, to securing and driving our cars and other means of transport, to checking and intervening in our physical and mental health during both waking and sleeping hours. In all probability, these robots will be linked up to a global network of other robots so that they can be updated with the latest scientific knowledge. In such a situation, it will be completely impossible for any human to have a complete overview of the complexity of these robots. As time goes on, these intelligent robots will create their own synthetic ethics.[1] In the initial phase, we can envisage the ethical behaviour of these robots as being based on the ethics they have been programmed to operate in accordance with. Subsequently, they will learn from their own experiences and also, via their global interconnectedness, from the experiences of other robots.

As we are all well aware, historically technological innovations have always triggered gloomy predictions. The 19th-century Luddites[2] that destroyed machinery, because they suspected that the machines would take their jobs, would very likely have joined in contemporary choruses of doom and attempted to break futuristic technological dreams. And even if modern Luddites do not destroy machines directly, they do create laws, regulations and directives that impede the development of artificial intelligence and intelligent robots. The problem is that in a situation of global competition, countries that fail to create such laws will be the technological winners.

When referring to an intelligent robot that is based on artificial intelligence, we refer to a machine that far exceeds human intellectual abilities in the realm of logical intelligence. Since such a robot will have an extremely high level of logical and rational intelligence, it will also be able to design other intelligent robots and produce them by using 3D printing technology. Like a phoenix, these 3D-printed robots will emerge fully formed, without any process of evolution. While humans have evolved over millions of years, intelligent robots are being created using yesterday's skills. One of the consequences of such a trend will most likely be the mass production of intelligent "techno-beings" with a level of logical-rational intelligence that far surpasses that of any human. This is the point at which we can envisage that these intelligent robots may begin to reflect on what should be done with unintelligent humans, who are serving as a ball-and-chain on their development. It is here that we must take science fiction seriously, because we may be at risk of waking up to find that we are in a world populated by intelligent robots whose logical and rational abilities far exceed those of humans, but who lack humans' emotional, relational, social and – despite their flaws – ethical skills and moral intelligence. It is against this background that we can say that "humans are underrated" (Colvin, 2018).

When referring to robot ethics in this chapter, we mean ethical norms and moral codes that are designed and implemented into intelligent robots (Abney, 2014: 35). Applying this definition of robot ethics means that we will not consider the ethics of the people who design robots or ethical theory in general.

If intelligent robots have all the logical and rational intelligence in the world, but only that, then we will have created technological sociopaths with an enormous capacity for intelligence on a global scale: in other words, an explosion of logical intelligence the like of which we have never experienced in human history. Accordingly, the objective of this chapter is to investigate the question: how can ethical reflection affect the development of artificial intelligence? To answer this general question, we have developed two sub-questions:

1. How can synthetic ethics affect the development of artificial intelligence?
2. How can robots capable of ethical reflection affect the development of artificial intelligence?

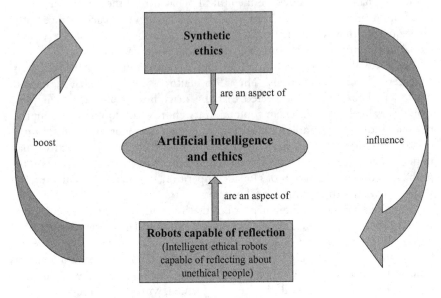

Figure 3.1 Artificial intelligence and robot ethics.

In addition, our objective is to enable us to cope adequately with techno-logical development and the innovation economy, so that neither modern-day Luddites nor technology-obsessed super-optimists get the opportunity to determine the course of society.

We have summarized this introduction in Figure 3.1, which also illustrates how we have structured this chapter.

Part I Synthetic ethics

We will examine the following question: how can synthetic ethics influence the development of artificial intelligence?

It seems reasonable to assume that in the future, machine intelligence will be integrated with artificial emotions, social intelligence and moral intelli-gence. Moral intelligence will be somewhat similar to what we might call syn-thetic ethics, that is, artificial ethics. These ethics are termed artificial, because they will be created by humans and implemented in robots. These synthetic ethics will be infallible in the sense that robots with integrated synthetic ethics will not be affected by historical, social and psychological motives and factors when making ethical choices, that is, factors which might affect humans. In situations where a robot has to make a decision involving ethical considerations, the robot will not be stressed or distracted, but evaluate each situation morally "correctly".

In the ethical sphere, the robot will evolve from being a tool and a substitute, from supporting human moral considerations, to taking over moral judgements and decisions when a situation arises, as when a self-driving car has to choose between injuring one of two people, such as an old man or a child, when a collision is unavoidable (mentioned in Chapter 2). The actions of these robots will be based on "pure synthetic" morals, which will not be influenced by psychological, historical or social factors. However, before we continue, and in order to avoid a misunderstanding, we would like to point out that the intelligent robots with synthetic moral intelligence will rarely be visible to us humans. They will be embedded in the things we use, and will increasingly become part of our everyday lives.[3] Consequently, these robots will not be created in our image, such as we often see in science fiction movies; thus, they will not live among us as avatars. The "invisible" robots will be embedded in our tools and machines such as kitchen appliances, self-driving cars, drones used in fishing and agriculture, surveillance drones policing law infringements, medical-diagnosis robots and surgical robots, as well as in combat drones used in armed conflicts. In addition, these intelligent robots will be interconnected globally with other similar robots. This interconnection between robots will greatly differentiate these informat-robots[4] from autonomous standalone robots.

The globally connected intelligent robots, which have developed synthetic ethics, will be able to make decisions continuously without us being aware of this. As mentioned, this may concern whether a self-driving car should choose between injuring a child or an old man in the event of an unavoidable accident. In a fraction of a second, even before the driver of the car manages to perceive the situation, the invisible intelligent robot in the car makes a decision and acts, based on the synthetic ethics implemented in the car's intelligent system. Synthetic ethics is just one component of the logical-rational intelligent robot, which is the car's "brain". However, this component is connected globally through various networks, so it can continuously learn from other synthetic moral machines around the world. Synthetic ethics will even be able to create moral innovations through learning and reflection.[5]

Most people would agree that it would represent progress if a self-driving car had a functioning synthetic ethics system that could take decisions in various hazardous situations, minimizing injury and damage. One could also perhaps consider the potential need for synthetic ethical intelligence to be introduced today in the case of intelligent banking systems that utilize investment algorithms, enabling them to make one-minute decisions that may have serious financial consequences for certain groups of people.

Description

For a long time, researchers have been fascinated by the relationship between machines, robots and ethics (Lin et al., 2012; Wallach & Allen, 2006; Anderson & Anderson, 2007). On the other hand, few researchers have turned the question on its head and asked how ethical reflection can affect the development of artificial intelligence on a general or specific basis (there are a few exceptions such as Dumouchel and Damiano [2017]). As a general rule, the preferred question has been how the development of artificial intelligence can affect ethical reflections. If we start by first considering the ethical dimension, before discussing matters in the engineer's machine room of how to develop artificial intelligence, then we will arrive at certain conclusions that would not have been apparent if we had asked the question the other way around, i.e., how artificial intelligence can affect ethical reflections.

If we first begin with ethical considerations before programming moral rules into intelligent robots, we will in the case of combat drones, medical robots, financial robots and the like also be programming limitations into the robots' action repertoires, which would not have been the case, if we had only reflected on ethical questions after the robot had calculated the most optimal action sequence.

Let us consider the combat drone and the financial robot as examples. Both of these robots will be freer to make the most logical-rational choices, if they do not have to consider the ethical consequences of various courses of action. Both of these examples show what forces would most probably come into play in order to resist putting ethics first on the agenda.

Ethical rules may be used as a criterion for a robot action, or a robot may act on the basis of an ethical norm that is based on certain principles of adequate ethical conduct. However, these represent two qualitatively different intervention procedures regarding robot decision-making. Let's take an example. The statement "you should not kill another person" is not a rule, but a religious and a human ethical norm. This norm provides some guidelines, but it is not absolute. On the other hand, a rule based on this statement, which is converted into an algorithm and implemented in a robot, provides a completely different sequence of actions from that of the norm. In this example, the norm would provide the robot with more flexibility in decision-making than the rule. The point of this example is to show that this is a distinction that will make a real difference to a sequence of actions when the ethical perspective is brought into the discussion regarding a robot's actions. It may well not be more effective in relation to specific goals to consider the ethical perspective before an action, but it would definitely lead to a more ethical pattern of action. The general and the businessman may point out that the most important thing is to optimize the best course of action concerning specific goals given the competition and the dangers with which one is faced. It is possible that they are right from their own standpoint, but this is not the same as saying that it is ethically right just because they are right; it may rather be the opposite. Being "right" is not necessarily the same as something being right. What is right for the businessman who uses a financial robot, disconnected from ethical reflections, is not necessarily right for the workers and the society affected by the robot's decisions. The theoretical point here is that ethical reflections should come to the forefront of the development of artificial intelligence. If ethical reflections come only after a robot action, the ethical perspective is likely to lose in the contest with the goal-effectiveness argument.

Robot ethics must not be confused with the term roboethics. "Roboethics" may be understood as referring to how people relate ethically to the use of robots. Robot ethics, on the other hand, addresses issues of the kind we are examining in this chapter and in this book overall, namely how intelligent robots should act ethically, or as expressed by Dumouchel and Damiano (2017: 172), "It is concerned with making robots moral machines by teaching them the difference between right and wrong".

Analysis

The development of robots over the past few years has been very successful in terms of increased productivity. We have witnessed this in the automotive industry, and in the healthcare sector, where the heavy lifting of patients is done by robots (Doudna, & Sternberg, 2018). However, it seems intelligent robots have been so effective that ethical considerations regarding the consequences of their use have not received much attention. Industrial robots and drones have contributed greatly to increased productivity and goal-effectiveness. In addition, robots have also been used in other sectors such as the entertainment industry; security systems; the fire and ambulance services; home-help and elderly care; warehouse systems and lifting of heavy goods; hospital diagnosis, surgical procedures, aiding the disabled; and as learning machines for children and adolescents in general, but especially for those with various types of learning difficulties.

These robots are efficient and goal-oriented, but have little or no ethical standards or rules built into their programming. It is when these robots reach a new "evolutionary" stage that it all starts to get complicated. It seems reasonable to assume that the new robots will have a high logical and rational intelligence. It also seems probable that the new intelligent robots will build further on the success of the robots that greatly increased productivity. If this assumption is correct, the ethical perspective will be pushed aside and, at best, taken into consideration only after completion of the design of the intelligent robot. In such a case, we will have designed a "sociopath" robot, that is, a robot with a high degree of logical intelligence, but without social, emotional or moral intelligence. It is such a development that is worrying, something that has concerned a number of people in related fields such as Stephen Hawking and Bill Gates (Floridi, 2017: 156).

It seems a reasonable assumption that most people support the use of intelligent robots to promote productivity, subject to certain conditions such as employment. Another condition is linked to emotional, social and moral intelligence. However, there is much to suggest that rationality will be given precedence and will push aside ethical considerations (Bekey, 2014: 17).

We are in the process of developing logically intelligent robots with no emotional and moral intelligence. Is this in fact the first step towards the entrance of the Four Horsemen of the Apocalypse?[6] To reduce the possibility of a future "apocalypse", emotional intelligence will be a necessary condition, so that intelligent robots are able to act with something more than just logical-rational intelligence. If there is one thing that emotional intelligence does better than anything else, it is to facilitate relationships and ties. Using synthetic ethics may be viewed as a sufficient condition if we are to move away from a future "apocalypse" that may happen as a result of our quest to develop robots that are solely logically intelligent.

Theoretical points

Synthetic ethics is not necessarily a poor substitute for human ethics. It may be the case that human ethical assessment will have greater authority, because it will be considered more autonomous than the synthetic ethical assessments of future intelligent robots. Autonomy seems to be a central concept in modern ethics, write Dumouchel and Damiano (2017: 173). Our point is that although autonomy is important in ethical assessments, autonomy does not guarantee morally correct behaviour. What is right and wrong cannot be specifically linked to autonomy, but instead to decisions and the consequences of decisions for living organisms and nature. Of course, teaching a robot the difference between right and wrong will not make the robot autonomous, but it may be the case that the decisions made by the robot will be more morally correct than those made by many autonomous humans, to the extent that one can say that humans are autonomous.

Implementing synthetic ethics in robots will have the aim of the robot behaving ethically correctly in various situations, despite the fact that it encounters autonomous people who use their ethical assessments. The point in this context is that we do not argue here that intelligent robots should make ethical assessments autonomously. On the contrary, it is our opinion that the moral actions of robots will be enhanced if they are connected to other intelligent robots globally. In such a context, the intelligent robots will be constantly updated in relation to the learning and experiences of other intelligent robots. It seems reasonable to assume that it is the moral actions that are of interest, not whether these actions are performed by autonomous humans or autonomous robots.

Proposition 1: Autonomous "agents" do not act more morally than interconnected agents, because it is the actual moral actions that are of importance.

Should we wait until we know what we have to deal with, before we start to discuss robot ethics? In other words, we do not know what we do not know and should therefore sit calmly in the boat until we have knowledge of what we don't yet know anything about. This way of thinking is relatively widespread, especially among engineers who design artificial intelligence and intelligent robots (Wallach & Allen, 2008; Tzafestas, 2016). The objection is relatively simple. Should we have waited until we saw the consequences of an ideology that likened the Jews to insects and vermin, before we intervened? Should we have waited until we saw the consequences of an ideology that looked at the Armenians in Turkey as cancerous tumours and parasites before we, as humanity, intervened? Yes, many did wait, and now history has shown us what happened. Should we have waited to argue against the Young Turks' genocidal ideology against the Armenians, before we intervened? Yes, most people did, and the result was the Armenian genocide. We could look at countless other examples throughout history. This is the main ethical objection to waiting until we see what results we have before we intervene.

Proposition 2: In the design and construction of intelligent robots, the norms and rules of synthetic ethics should be drawn up and integrated within the robot, before the installation of logical and rational artificial intelligence, because of the danger of engineers closing the window on the integration of ethical intelligence into the robot, if the logical and rational artificial intelligence is installed first.

Of great interest is the development of intelligent robots that are designed to achieve clear goals, while also incorporating norms and rules in their design that prevent or limit them from achieving these goals, because it would be contrary to moral rules. What should the robot do? Should it under no circumstances ignore the moral rules, or should the ethical requirements be viewed in light of what can be achieved for many if certain goals are reached? There are already robots that act on the basis of set goals, but which are not equipped with built in moral rules. For example, the ethical implications of the actions of military drones are assessed by military personnel. Soldiers, however, have to follow the chain of command, and will only break the principle of obedience in the rarest of cases.

Proposition 3: Intelligent robots should be equipped with built in moral rules that override the specific goals for which the robot is designed.

Practical utility

In this section, we will present five possible practical benefits of how synthetic ethics can affect the development of artificial intelligence.

1. Synthetic ethics and legal aspects
   The first practical advantage we will look at relates to legal aspects. If an intelligent robot equipped with moral intelligence is installed in a car, a boat and the like, then it is probable that insurance companies will offer lower premiums than if moral intelligence was not installed. In road traffic, moral intelligence linked to the legal system could make a great impact. There are many instances where artificial intelligence could be used to override the illegal or hazardous actions of a driver; to mention a few: exceeding speed limits, driving through a red light, talking on the phone while driving, driving in a state of intoxication, driving while tired, unwell, etc. Cars equipped with moral intelligence will be able to prevent or reduce these and other illegal and hazardous actions of a driver, although not all.

2. Synthetic ethics and psychological regularities

There are few, if any, psychological laws that specifically affect behaviour. On the other hand, there are countless regularities or social mechanisms that say something about the relationship between psychology and behaviour. The point is that even for humans, there are no clear relationships between a psychological state and behaviour.

For a logical-rational intelligent robot, on the other hand, there is a clear goal and means relationship. For such a robot, the goal justifies the means, because this is how the robot is designed. However, this development may result in us letting loose the horsemen of the apocalypse. In order to prevent such a catastrophe, one can integrate synthetic ethics into intelligent robots, thereby developing moral intelligence, which in every context assesses the goal and means relationship from an ethical and moral standpoint.

3. Synthetic ethics and "requisite variety identity"

Synthetic ethics will ensure that an intelligent robot is perceived as a helpful tool by an assistant or the like, which humans can then rely on in various contexts and situations. Thus, it is not only an issue of the efficiency and productivity that will be the robot's function, but the question also arises of whether the goals and methods used to achieve goals are morally justifiable. If intelligent robots are developed along these lines, it seems reasonable to assume that most people will view artificial intelligence as a positive technology, which can help humanity achieve goals without compromising ethical principles. It is in such a context we can say that the intelligent robot has an identity with sufficient requisite variety.

4. Synthetic ethics and efficiency

Let's perform a thought experiment: a driverless bus travels from A to B every day, 24/7 for ten years without a single accident. One day, however, it happens that the bus, instead of turning left at a junction, drives straight ahead and into a crowd of people. Many people are killed and more are injured for life. The bus had worked perfectly for ten years, but this one accident raises the question of whether it is safe to use driverless buses. Let's assume a statistician puts some facts on the table. He demonstrates that when the buses were driven by humans, then over a corresponding distance and time period more people were killed and injured than in the case of the single accident with the driverless bus. However, those killed and injured were spread over several years and did not receive the same media attention.

However, the frenzied reporting of the media should not form the basis of the assessment of artificial intelligence; instead, the basis should be an objective study of what is best for most people. In this thought experiment, it was shown that using regular drivers resulted in accidents with more injuries and deaths than using robots as drivers. Let us consider a similar situation with surgeons and surgical machines. If a robot was responsible for diagnosis and surgical interventions for a period of ten years and every-thing went well, until after ten years there was an error that led to a death, should one stop using robots for surgical interventions? Similar to the pre-vious thought experiment with the bus, one would have to compare this with a situation where only human surgeons did the interventions. If the comparison came out in favour of the surgeons, then one would have to carry out a larger analysis of what is most effective. If the larger analysis fell in favour of the robots, then one should continue to develop and apply artificial intelligence in surgical procedures.

5. Synthetic ethics and values

If the technological development of intelligent robots and artificial intel-ligence continues at the same rate as today, and the focus on synthetic ethics also continues at the same rate as today, then a huge gap will develop between intelligent robots/artificial intelligence and synthetic ethics. If society allows this gap to exist and increase, it seems reasonable to assume that society will change its norms and values, beliefs and ultimately its ethics. Based on historical examples, the assumption here is that technology is usually given priority, and ethical values become plastic and change according to what serves technological development best.

Sub-conclusion

We have examined the following question in this section: how can synthetic ethics influence the development of artificial intelligence?

The simple answer is that synthetic ethics can influence the development of artificial intelligence, so that the scope of action of rational-logical intelligent robots will be curbed. In short, goal-efficiency will be reduced, while the eth-ical quality of actions will improve.

However, there will be a great number of ethical challenges that will arise when developing artificial intelligence, which cannot be resolved immediately by implementing synthetic ethics in intelligent robots. For instance, can the algorithms used in robots be discriminatory?

Basically, these algorithms should be objective and not coloured by discrim-inatory attitudes. For instance, algorithms may be used by a bank to assess if a loan applicant should be granted a bank loan. The algorithm could process various pieces of information relating to the applicant, such as residence address, how many times he/she has changed residence over the last five years, previous income, etc. However, the algorithm would not necessarily take into account other variables such as current or future opportunities. Thus, a "neutral and objective" algorithm can make choices based on subjective perceptions that are included in the algorithm. It must therefore be possible to assess and check the algorithms that are developed and possibly change them so that the technology is not given priority over ethics.

Part II Reflective robots

In this part, we will examine the following question: how can ethically reflective robots affect the development of artificial intelligence?

Once singularity has occurred in around 2035–2040 (Chace, 2016; Shanahan, 2015; Kurtzweil, 2005; Johannessen, 2020), we can expect that intelligent robots will have the capacity to reflect on their actions, as well as on the ethical code upon which these actions will be based. However, intelligence, ethics and reflection are not necessarily related concepts. Both humans and machines may have high logical-rational intelligence, without being particularly ethical in their behaviour or reflective in their thinking. It is when the three elements interact that we can speak of reflective robots. The working definition of a reflective robot that we use here is as follows: intelligent robots with a high degree of logical-rational intelligence, a high degree of synthetic ethical awareness and a high degree of synthetic reflective ability.

In order to avoid an unproductive discussion about whether intelligent robots will ever be able to reflect in the same way as human beings, we introduce here the concept of synthetic reflection, which is shown in Figure 3.2. An intelligent robot uses synthetic reflection when it makes use of logical-rational intelligence and synthetic ethics in order to make choices between different options of action. With such a straightforward definition, we will avoid associating reflective robots with something chilling and science-fiction-like. Synthetic reflective ability may also be used to reflect on whether a decision made by a human being actually attained the desired result, and what should have been done instead to improve the result. Synthetic reflection can thus provide a greater degree of requisite variety for decision makers.

Figure 3.2 illustrates what is meant by reflective robots.

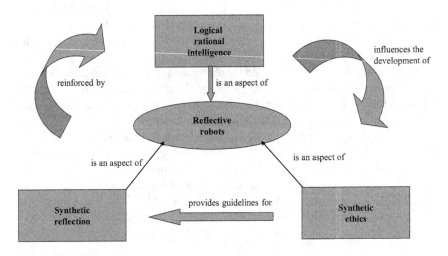

Figure 3.2 Reflective robots.

Description

Why should we investigate the question about reflective robots and ethics now, when we know that it is far ahead in the future that robots will have the ability to reflect intelligently?

A day doesn't go past without the media reporting something new about robots, automation, unemployment attributed to AI and robots, and entire functions in production being taken over by robots. But we hear little about robots having the ability to reflect intelligently, so why should we focus on this now? The answer to this question is that it is important for society to be prepared to tackle this situation when it arises. Moreover, if we don't start to focus on future ethical challenges, then it is most probable that technological and economic rationality will be given priority.

Should we fear the development that most people working with technology see coming? We should certainly be concerned. Imagine you are travelling by ship, and the radar detects an iceberg, but you don't react because it's in the far distance, so you wait before doing anything. The point is that we don't have control of the weather, wind and currents in the ocean, so the iceberg may turn out to be a danger in a few hours' time, and not, as we believed, much later. This may also be the case with reflective robots and ethics. If, at an early stage, we consider the various scenarios related to reflective robots, artificial intelligence and ethics, then we will be better prepared if and when such a situation occurs. The worst-case scenario is that we leave developments to technological and economic rationality.

John McCarthy, the computer analyst, who coined the term artificial intelligence in 1955,[7] remarked that as soon as AI was used in practical contexts it was called something else (Vardi, 2012: 5). Similarly, in the not too distant future, when we will have reflective robots, it is also probable that no one will call them robots anymore. Once we have developed reflective robots, the technology will be so complex that it will be impossible to understand exactly where the ethical decisions are made in the reflective process. Some of the decision units in these robots are likely to be so small and complicated that we will not be able to see them with the naked eye, and they will be so complex that even if we did see them, we would not comprehend how the ethical element decided on one action or another.

This is one of the main reasons why we should discuss and find solutions to the ethical challenges and problems, long before reflective robots become a reality. It is now we have just such an opportunity, so we should intervene now and thereby influence and manage developments. Our point here is that if we wait until technological development has created these reflective robots, it will then be too late.

Whether artificial intelligence and intelligent robots will be able to possess consciousness, and whether they will have the ability of reflective thinking, is dependent on what we mean by the term "artificial intelligence". Robot arms with artificial intelligence carry out production processes in modern automobile factories that are far more efficient than human workers. Spacecrafts launched towards the Moon or Mars are controlled by artificial intelligence. The point here is that artificial intelligence is already embedded in many of our modern technological devices. However, whether this artificial intelligence will ever be able to reflect on its altogether own behaviour and possess consciousness is another matter. This is also dependent, however, on how we define reflective thinking and consciousness.

Manipulation of large amounts of data so that a pattern can be uncovered is certainly a task that intelligent robots can already perform today. There are many enterprises, institutions and organizations that are currently working on the challenges associated with artificial intelligence.[8]

The consideration of ethics is often viewed as a limiting factor in the development of artificial intelligence (Boddington, 2017: 7). However, taking ethical questions into consideration when designing artificial intelligence will reduce the uncertainty and fear that most people associate with its development.

The key question is whether intelligent reflective robots are a real possibility in the future, and how the consideration of ethics can influence their development. Robots that possess consciousness and have the capacity for reflective thinking may be a long way off in the future; nevertheless, regardless of if and when this occurs, we should already now be considering what impact this can have for individuals, organizations and society.

Analysis

Artificial intelligence is used in many contexts such as in the composition of music and in the development of complex algorithms that are used to reach public policy decisions (Boddington, 2017: 2). Artificial intelligence is also used to a considerable extent in medical equipment, the diagnosis of patients,[9] to aid and support surgical procedures and even in therapy procedures.[10] It is also used to manipulate huge amounts of data in order to reveal patterns that can determine the spread of diseases and epidemics. Intelligent robots have been used in Japan in the care of the elderly, as well as in the nursing of sick elderly patients. Intelligent robots are used to stimulate patients suffering from dementia. Artificial intelligence is also used in the development and use of artificial limbs. Autonomous cars and automatic pilots in aircraft and boats also use artificial intelligence to some extent.[11] All the devices and everyday objects that go under the name of "the internet of things" use artificial intelligence for some or all of their operations. In commerce, stock market trading that was previously done by people has been replaced to a large degree by artificial intelligence.[12] The same applies to the sorting and pricing of airline tickets that have now to a great extent been automated.

Taking the examples above, we can see that artificial intelligence has already been brought into use today in many different contexts.

Case study 1

Let us suppose that the driver of a semi-autonomous car suddenly becomes unwell and loses consciousness. At the time, the car is travelling along a narrow road, with pedestrians ahead and on either side. The car is travelling too fast to make an emergency stop. Peter, who is parked in a driveway in his breakdown truck, sees the situation and has a chance to alter the car's course. If he takes this opportunity, he will save the lives of five or six people who are walking along the road, but the car will very likely kill a child, who is sitting playing on a side road. The question is whether it is ethical for Peter to choose to prevent the deaths of five to six people, even if the child is likely to be seriously injured or even killed. This is a classic ethical dilemma and was first proposed by Thomas Aquinas in the 13th century (Mikhail, 2007: 143–152). Empirical studies show that most people would think the ethical choice is to save the larger number of people at the expense of an individual or smaller number of people.

Case study 2

Continuing from case study 1, the following happens: Per pushes a heavily built woman in front of the car so that it is brought to a halt. The woman is killed, but the five or six people survive. Is it morally correct for Per to push the heavy-built woman in front of the car to save several people? Empirical studies show that it is not deemed morally acceptable for Per to push the woman in front of the car, even though several people are saved (Hauser et al., 2007: 1–21). So it seems that such a direct action is not morally acceptable, such as pushing the woman in front of the car, but we are more willing to accept an action that is more indirect, where many people are saved, and where the negative consequences can be considered as being a side effect.

Case study 3

Following on from case studies 2 and 3: the semi-autonomous car is unable to stop without someone being injured. In this case, Per faces another challenging decision. He can throw a heavy object onto the road, so that the car's course changes and enters a side road where it hits a man who is killed. Should Per carry out this action?

Empirical studies show that in this case, 56% respond (Hauser et al., 2007: 1–21) that it is defensible to throw an object in front of the car so that it changes its course, even if this results in a man being killed. It seems that when there is no intention to injure someone, most people find it morally correct to intervene in this way.

Case study 4

Let us carry out a small thought experiment: if we develop intelligent machines with an exceptionally high degree of logical and rational competence, but absolutely no emotional, social or ethical competence, how will these intelligent robots reflect and act in relation to the people who created them, almost in their own image? Will they become submissive and carry out the orders they are given by their human "masters"? Will they become autonomous and decide themselves what is best for humans? Of course, no one has the answer to these questions. Therefore, we can rather ask the question: what will benefit the logical-rational robots? The answer seems to be quite clear. The intelligent robots will most likely find it expedient to use humans as an aid to supply them with the resources they need for their evolutionary development, which will probably be the integration of emotional, social and moral intelligence in their design. Intelligent robots and the evolutionary development that these robots will undergo are aspects of the concept of singularity. This will eventually result in an explosive cascade of intelligent robots that will penetrate every function of our lives, being the future reality of which we already see the contours today. On the other hand, there is little focus on the integration of emotional, social and moral intelligence into these intelligent machines. How will humanity survive if other "techno-beings" can, in a more rational manner, safeguard the sustainability of the planet Tellus? Just as we have used animals and nature to provide us with our needs, such as food and shelter, intelligent robots may well use people as their "servants" to supply them with what they need, in order to evolve from logical-rational "techno-beings" into "techno-beings" with an integrated moral, social and emotional intelligence (Johannessen, 2020). When this evolutionary development is complete, what use will the "techno-beings" have for unintelligent people?

Stephen Hawking, Elon Musk and Bill Gates are all very wary of the developments that intelligent robots can undergo (Floridi, 2017: 156). These three men having worked in related fields are all more than qualified to express their views on this subject. Given this background, the thought experiment above presents a not an entirely unthinkable scenario, although today it may seem more like science fiction. However, in this context, it should be noted that there are a lot of hard facts in science fiction and a lot of soft facts in science. The possible but unlikely can quickly become a reality with which we all have to deal.

Theoretical points

According to Scanlon's contractualism, an act is only immoral if it violates the fundamental principles of the context in which the act takes place, even if it is morally acceptable in another context (Scanlon, 1998). This sounds acceptable at first glance, and calls to mind the proverb, "When in Rome do as the Romans do". However, if we were to take the meaning of the proverb literally, we might ask which Romans are we talking about? In Rome there were different codes of behaviour depending on which social class one belonged to. If you were in Rome, what would be the "correct" moral code to choose? In Scanlon's view, homosexual behaviour may be viewed as morally wrong in a narrow sense within a specific context. However, in the same but broader context, homosexuality would be viewed as a legitimate moral sexuality. Thus, Scanlon's contractualism does not provide any real guidelines for action, but is merely an academic exercise aimed at promoting a certain kind of morality in a particular context, without considering the various sub-contexts. Scanlon's approach is clearly concerned with showing that moral actions can be flexible. An important point, however, is that contractualism enables such a broad scope of interpretation that it does not provide much meaning for the individual. "When in Rome, do as the Romans do". Which Romans are we talking about? The elite and their morality, or plebeian morality? If the morality of the elite permits murder, should one also be able to commit murder when "in Rome", while adhering to a moral code in the same context?

The development and use of artificial intelligence present clear practical challenges to moral problems in everyday life. First, this concerns the development of artificial intelligence. As a rule, those who develop artificial intelligence are not trained in ethical theory and reasoning; they are mainly skilled engineers who design and program various types of algorithms in robots' software. Once these algorithms have been programmed into the intelligent robot, it will be difficult, if not impossible, to integrate moral intelligence into the same robot afterwards. In addition, the financial resources devoted to the research of artificial intelligence are mainly directed at the logical-rational intelligence of these robots, and not so much at the ethical challenges and problems that may result from artificial intelligence (Bastos de Morais, 2014: 11). When the greater part of research funding is allocated to the development of artificial intelligence without taking ethical issues into consideration, this will then result in an imbalance, where ethical issues are given little attention. The focus on artificial intelligence will largely lead to those professions that design artificial intelligence gaining more power and authority, in terms of research, than those who are concerned with the ethical challenges and issues.

The arguments stated above indicate that we should develop a policy at the political level that ensures that ethics becomes an integral part of the research on artificial intelligence and intelligent robots. This should be done at an early stage, before singularity occurs (Bastos de Morais, 2014: 11) and when intelligent robots are developed and become a reality. Even if super human intelligence is not developed, or singularity does not occur in the near future, it is nonetheless important for ethics to be integrated into the design of intelligent robots and artificial intelligence at the very beginning of the process.

It is highly probable that when artificial intelligence is connected to the sensor systems that connect the intelligent robots to information and knowledge processes in the global environment, we will then witness super human intelligence evolving in the intelligent robots. At this point in time, well into the future, these robots will develop the ability to reflect upon their decisions and learn from their actions; it is at this point that we will witness the outlines of what can be called "the global brain".

The ethical dimension should be a natural part of the development and design of artificial intelligence, so that we are prepared in advance of new developments. Unfortunately, ethical challenges and issues related to artificial intelligence are rarely given much attention in the public debate (Bastos de Morais, 2014: 11). Innovations in general, and in artificial intelligence in particular, are developed without any particular consideration of their ethical consequences. As a general rule, it is only in the wake of actions related to artificial intelligence that the ethical consequences become problematic.

Technological developments in general, and in the development of artificial intelligence in particular, are impossible to predict. If we wait until the technological changes have taken place before we consider the ethical challenges related to the actions of intelligent robots, it may then be too late. Technology related to artificial intelligence is developed in various contexts, ranging from the construction of combat drones to surgical procedures. The widespread use of this technology requires us to take into account both context and complexity when discussing ethical issues and implementing ethical guidelines (Boddington, 2017: 67). We also know that technological changes are closely linked to organizational and social changes. It is therefore crucial that we are in advance of technological development regarding the consideration of ethical issues.

Several ethical issues will arise when the use of intelligent robots and artificial intelligence becomes common in working life. Working life will undergo great changes, and most jobs will be in danger of being completely or partly taken over by rational, intelligent robots. Amongst other things, when employees lose their jobs because their work tasks are taken over, this will result in great ethical challenges. Does work have a value in itself? Do people need meaningful jobs? Should society develop ethical rules in advance of the development of artificial intelligence, or should ethical issues be considered only after problems occur in relation to the use of technology? The most important question, which we consider crucial on this, is whether ethical issues and guidelines should receive as much attention as new technological developments. How can we ensure that ethical rules and guidelines are not subordinated to the development of new technology and the profit that is created from technological development?

Proposition 1: Ethical reflection should take place in advance of robotization in general, and of artificial intelligence in particular.

Should the debate on ethics and artificial intelligence be linked to an evolutionary development of moral rules, in relation to artificial intelligence? If the answer to this question is yes, it is then not so imperative to develop ethical guidelines in advance of technological developments. Many believe the benefit of such an approach is that technological development will progress more rapidly, because we will not have to take into account ethical challenges in advance of technological developments. Another objection is that it will be impossible to discuss ethical challenges in advance of technological developments, because the development of technology cannot be predicted, but is the result of interacting complexity that creates emergents. Evolutionary moral development will, it is claimed, create balancing moral norms when we see how technology evolves (Pereira & Saptawijaya, 2016: 7). The idea is that as technological development progresses, various ethical dilemmas will arise, which will then be solved, with the solutions implemented in the technological systems. In this way, the ethical dimension of the intelligent robots will develop evolutionarily in relation to practical contexts. It is not necessarily the "truth" that is essential in this perspective, but what is good in practice for individuals, organizations and society, as a result of synthetic ethics being implemented after technological developments have become apparent. One can call this perspective reflective synthetic ethics. This reflective synthetic ethics will be evolutionary and will emerge in the wake of technological developments. One could rephrase the proverb that "hindsight is the only true wisdom", with the statement that reflective ethics is the only sustainable ethics, because it is based on closeness to actual context and practices.

Proposition 2: Reflective synthetic ethics is closely related to the practical context, and will be easier to implement, and accepted by those who are affected by the consequences of the use of artificial intelligence and intelligent robots.

If we assume that the artificial intelligence we have developed can learn from its own actions and mistakes, then this artificial intelligence will quickly become much better than the intelligence we have developed ourselves. One can imagine that a semi-autonomous car chooses to run into a man in order to avoid colliding with two women. The women were walking in the road in the path of the car while the man was walking along the pavement.

If an empirical study showed that most people believed that the women were themselves to blame because they were walking in the road, and the semi-autonomous car should have chosen to run into them, and not into the man who was walking along the pavement, then the car will have "learnt" something that it could use to make a decision concerning a similar event in the future. Through its internet connection, the car will also communicate this information to all other cars that are connected to the internet. In this way, all the cars will have "learnt" a new moral rule of action.

Proposition 3: In the case of reflective synthetic ethics, both intention and consequences must be considered, before we can say whether an action is morally correct.

Practical utility

In this section, we will present three possible practical benefits of how reflective synthetic ethics can influence the development of artificial intelligence.

1. The abstract versus the concrete
    There will always be tension in ethics between abstract thinking, which is concerned with general principles, and concrete thinking, based on a case-by-case assessment.

As with all abstraction processes, one will lose valuable information in the ethical abstraction process, because one needs to eliminate some details for the purpose of showing the larger picture. In this abstraction process, one therefore loses information that could be valuable in relation to an individual case. If the reflective robot bases its actions on information gathered from individual cases, it will have difficulty assessing what to do when an innovative new case arises. It therefore seems logical that there should be a connection between abstract ethical rules and concrete cases. The question is: how should this be resolved in practice. For the reflective robot, this may be resolved if the designer starts with some general principles that are culture-dependent, i.e., these general principles are not universal, but linked to the culture in which the robot operates. Furthermore, the reflective robot will apply these general principles within a context, i.e., the context will send certain signals that can affect the general principles. In addition, there will be some very specific cases within the context that further affect both the context and the general principles. In this way, the reflective robot will, by means of its reflective synthetic ethics, rely on all three elements: the general principles, the context and the individual concrete cases. We have illustrated this in Figure 3.3.

2. Intelligence versus ethics

A great deal of attention is focused on artificial intelligence; however, there is less focus on ethics in relation to artificial intelligence. The emphasis on artificial intelligence suggests that intelligence is humanity's most important characteristic. If such a view becomes established, then intelligence will be viewed as being more important than empathy, ethics and emotional competence. If such a situation continues, we may risk emphasizing

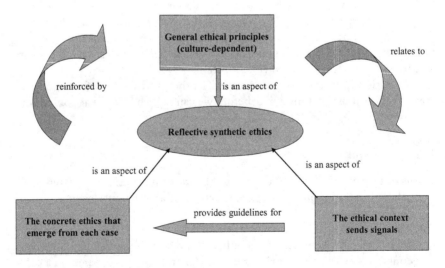

Figure 3.3 Reflective synthetic ethics.

logical-rational intelligence at the expense of ethical reflection. Another consequence of the emphasis on logical-rational intelligence is that intelligent robots will ultimately be valued as being worth more than people who do not have such as high an IQ as an intelligent robot. It is highly probable that in the near future intelligent robots will have a higher IQ than that of even a genius (Bekey, 2014). While evolution is often considered to be a product of survival of the fittest and of blind chance, the new development, with its targeted focus on artificial intelligence, can alter this view. If, on the other hand, we are able to focus attention on the role of ethics in general and of synthetic ethics in particular, we can then influence the development of artificial intelligence.

If we consider the evolution of humanity from the Stone Age to the present day, there is a strong tendency to focus on the development of human intelligence, and especially on how intelligence has enabled humans to develop technological tools to increase productivity. However, we know that there are other factors that have played a large role in evolution such as selection of mate, our sociality and our ability to organize (Barrett et al., 2002; Morgan, 2011). Is it possible that the ethics we reflect on in relation to the development of artificial intelligence may be understood as a new step in human evolution, that is, are we starting to consider ethical understanding as being more important than logical-rational intelligence?

As a result of the description above, synthetic ethics will have the potential to influence artificial intelligence in the following ways:

a. It may change the current emphasis on intelligence as a decisive factor in human evolutionary development.
b. It may completely change the design and use of artificial intelligence in relation to consequences such as the consequences of using combat drones.
c. It may change the emphasis on what is viewed as being important in human evolution: logical-rational intelligence or emotional, social and moral intelligence.
d. It may change our understanding of what is meant by work.
e. It may change our understanding of what is meant by "the good life".
f. It may change our understanding of what is meant by important work and what work is viewed as less important.
g. It will have the potential to give more emphasis to the importance of emotional and social competence.

3. Moral actions for humans and machines
    Moral behaviour and actions may be said to be different for humans and machines. In other words, one expects people to possess moral codes and act morally, but one does not necessarily have the same expectations towards a machine. Thus, it is expected that humans will have a greater degree of moral reflection than machines. If this assumption is correct, it may result in ethical issues not being taken into consideration to such an extent in the development of artificial intelligence and intelligent robots,

because we expect the people responsible for the robots to make the ethical decisions, rather than the robots. However, it seems reasonable to assume that autonomous cars, or perhaps even semi-autonomous cars, will be safer in the future than human-driven cars, for the simple reason that they will not be accepted if they are not safer.

The question of whether these cars can make better moral decisions than a human driver is a question of how well the designers have implemented ethical rules in the car's decision-making system. One can also imagine the designer only being concerned with technological performance, and that an ethical assessment unit is developed afterwards, which is then connected to the car's autonomous decision-making system. In this way, the ethics unit will not be something that is developed by the engineer and the software designer; this task will be given to those who have expertise in ethical decision-making.

No matter how good the self-driving cars are at making moral decisions, and no matter how many people are saved as a result of this technology, we face a major problem, which Boddington (2017) have expressed in the following way: "if self-driving cars cut the roughly 40,000 annual US traffic fatalities in half, the car makers might get, not 20,000 thank-you notes, but 20,000 lawsuits".[13] It is this assumption that should be incorporated into the decision-making system of self-driving and semi-self-driving cars, because it seems people are less tolerant of errors committed by machines than of errors committed by people.

Sub-conclusion

In Part II, we have examined the following question: how can ethically reflective robots affect the development of artificial intelligence? By building ethical behaviour into the decision-making system of intelligent robots, the development of artificial intelligence may be influenced in the following ways:

1. Strategy: When the designers of artificial intelligence must, as an absolute requirement, take ethical issues into account, or else be in communication with a team of people considering the various ethical consequences of the use of artificial intelligence, it seems reasonable to assume that it will take a longer time to develop logical-rational intelligence. The reasoning is simple: the more people there are who have something to say about the various assessments that the robot should make in ethical situations, the more time it will take before the designers can proceed to the next step in the development process. However, this explanation will be less relevant if the ethical component of artificial intelligence is developed separately and is then physically coupled to the machine at a later point.
2. Safety: It seems reasonable to assume that one important goal of developing artificial intelligence is that the functions to be carried out by the machines, within which the technology will be implemented, will be performed safely,

or more safely than if the functions were performed by a human being. However, the question that then arises is: safer for whom? If this concerns an autonomous car, we then have to assume that the person sitting in the car is chiefly interested in the car not harming him/her and the other passengers in the car. If this assumption is correct, and is something that can be investigated empirically, then the next question is: would the driver and the passengers feel comfortable about the fact that the autonomous car included an ethical assessment program for making choices in accident situations, which could result in the car making a decision that would risk injury to the car's occupants, in order to avoid injuring other people in the car's environment. If the answer to this question is no, then the ethical assessments that will be implemented in the car's decision-making system will be less complicated than if the answer is yes, and would therefore probably take less time to implement. We assume that this argumentation would be generalized as being relevant for other artificial intelligent systems such as combat drones and surgical robots.

3. Disaster ethics: Who should decide what is least ethically wrong: to run over and kill a three-month-old baby or drive into a group of senior citizens killing and injuring them? For the ethically reflective robot that is installed into, for example, an autonomous or a semi-autonomous car, this could be a real situation. Of course, it is not the reflective robot that creates the ethical rules installed in the car. The robot only carries out the instructions designed into its program. Most probably, a project group would develop the programs for these robots, and they would have to consider questions related to quantitative ethics. What is a "disaster" in an ethical context and who will decide what to do when a situation is defined as an ethical disaster?

Conclusion

The main issue explored in this chapter has been as follows: how can ethical reflection affect the development of artificial intelligence? The short and simple answer is that ethical reflection will add a greater degree of complexity to the development of artificial intelligence. Accordingly, more time will need to be devoted to the development of artificial intelligence. The other side of the coin is that if we spend longer on ethical reflection, we can expect that the decisions made by an intelligent robot in different contexts and situations will be based on a better foundation of ethical knowledge. It is possible for us to prevent ethical considerations from impeding the development of artificial intelligence. We can do this by ensuring that designers build interfaces that can be connected to ethical components. Such an ethical component could then be developed by a team that is separate from the team responsible for the development of artificial intelligence.

A significant point regarding the ethics with which robots and artificial intelligence are equipped is that it has little value if it does not have the full support of the social systems; for example, the organizations or institutions that will put it into use. One could say that an organization's ethical integrity has to correspond to the ethics being applied by the intelligent robots. If the organization's ethics is merely a form of window-dressing, then it is less significant if the applied ethics used by the intelligent robots deviates from the organization's stated ethical position. On the other hand, if an organization takes ethical integrity seriously, and gives it its full support, it is then crucial that the ethics applied by the intelligent robots corresponds to the organization's stated ethical position.

The theoretical point that we can deduce from the above arguments is as follows: the greater the degree of an organization's ethical integrity, the less discrepancy there will be between applied AI ethics and the organization's stated ethical position. One of the consequences of this theoretical point is the answer to the main question considered in this chapter: the process of ethical reflection means that the development of artificial intelligence proceeds more slowly than it would in the absence of such reflections.

The practical implications of this theoretical point are that, in any project where artificial intelligence is developed, it will be a requirement that an ethical council should follow up the development of this artificial intelligence. Such an ethical council should be composed that diverse ethical views are represented. The ethical discussions should be transparent, so that they can be critiqued. We acknowledge that such a process of ethical reflection will take time, slowing the progress of all projects related to artificial intelligence. Given the large-scale global competition that exists around the development of artificial intelligence, we are of the opinion that this process of ethical reflection will be sidelined, simply becoming a modern form of window-dressing, in order to show that account is being taken of the ethical perspective.

Some suggestions as to how one could continue investigations into ethics and artificial intelligence are as follows:

1. The ontological perspective: Commence large-scale empirical projects that aim to find out what people think about various ethical challenges at the intersection between artificial intelligence and ethics. This could be achieved by presenting respondents with different scenarios in relation to a specific situation.
2. Epistemological perspective: Develop more projects where different ethical perspectives[14] arrive, through a process of reflection, at solutions to specific cases. In other words, we are interested here in the different potential ethical outcomes to the various knowledge perspectives.
3. Utility-based perspective. Develop several projects, with the aim of revealing how developments in artificial intelligence can be advantageous, sustainable and responsible for the system(s) to which they will be applied. The purpose of such projects could be to diversify the ethical perspectives, such that some kinds of artificial intelligence could require a higher degree

of ethical reflection, while other projects would not require such extensive ethical reflection.

4. Develop a system, whereby there are two groups in every project working on artificial intelligence. One group would deal with the technological aspects and design a "port" for connecting the results of ethical decision-making processes. The other group would work on ethical considerations relating to the practical field in which the robot would operate. This ethical group would develop an ethical "junction box" that would be connected to the port installed by the technological group.

Notes

1 In this chapter, we use the term synthetic ethics in order to distinguish robot ethics from human morality and ethics.
2 https://en.wikipedia.org/wiki/Luddite.
3 Often called the "Internet of things".
4 Informat-robots are intelligent robots that are interconnected with other intelligent robots globally.
5 Intelligent robots that reflect on moral issues do so by comparing the outcome of decisions in relation to goals and evaluations made by, among others, people after an accident or an event. The intelligent robot can then use these decisions when making future assessments and decisions. It is this kind of reflection we are talking about here.
6 The Four Horsemen of the Apocalypse are described in John's Book of Revelation, chapter 6, verses 1–9.
7 https://en.wikipedia.org/wiki/John_McCarthy_(computer_scientist).
8 We have mentioned some practical examples in the analysis section.
9 See the following website for research on AI, and the consequences for society: www. partnershiponai.org.
10 The Leverhulme Centre for the future of intelligence at Cambridge: http://lcfi. ac.uk.
11 The One Hundred Year Study on Artificial Intelligence (AI100) at Stanford University: https://ai100.stanford.edu.
12 The Machine Intelligence Research Institute (MIRI): https://intelligence.org.
13 Cited in Boddington, 2017: 88.
14 Examples of these different ethical perspectives could be virtue ethics, consequentialism, proximity ethics, intentionalism and so on.

Bibliography

Abney, K. (2014). Robotics, ethical theory, and metaethics: A guide for the perplexed, in Lin, P.; Abney, K. & Bekey, G.A. (eds.). Robot ethics: The ethical and social implications of robotics, MIT Press, Cambridge, MA. pp. 35–54.

Allen, C.; Wallach, W. & Smit, I. (2006). Why machine ethics? IEEE Intelligent Systems, 21, 4: 12–17.

Anderson, M. & Anderson, S.L. (2007). Machine ethics: Creating the ethical intelligent agent, AI Magazine, 28, 4: 15–26.

Barrett, L.; Dunbar, R. & Lycett, J. (2002). Human evolutionary psychology, Princeton University Press, Princeton, NJ.

Bastos de Morais, J-C. (2014). Artificial intelligence: Innovation with an ethical choice, in Bastos de Morais, J-C. & Stückelberger, C. (eds.). Innovation ethics, Globaletics. net, Geneva. pp. 11–17.

Bekey, G.A. (2014). Current trends in robotics: Technology and ethics, in Lin, P.; Abney, K. & Bekey, G.A. (eds.). Robot ethics: The ethical and social implications of robotics, MIT Press, Cambridge, MA. pp. 17–34.

Boddington, P. (2017). Towards a code of ethics for artificial intelligence, Springer, London.

Boden, M.A. (2016). AI: Its nature and future, Oxford University Press, Oxford.

Chace, C. (2016). The economic singularity, Three Cs Publishing, New York.

Colvin, G. (2018). Humans are underrated: What high achievers know that brilliant machines never will, Nicolas Brealey Publishing, New York.

Doudna, J. & Sternberg, A. (2018). A crack in creation: The new power to control evolution, Vintage, New York.

Dumouchel, P. & Damiano, L. (2017). Living with robots, Harvard University Press, Boston, MA.

Floridi, L. (2017). The ethics of artificial intelligence, in Franklin, D. (Ed.). Mega tech, technology in 2050, Profile Books, New York. pp. 155–163.

Franklin, D. (2017). Mega tech, The Economist Books, New York.

Hauser, M.; Cushman, F.; Young, F.; Jin, L. & Mikhail, R.K. (2007). A dissociation between moral judgments and justifications, Mind & Language, 22, 1: 1–21.

Johannessen, J-A. (2020). Singularity innovation, Routledge, London.

Kurzweil, R. (2005). The singularity is near, Penguin, London.

Lin, P.; Abney, K. & Bekey, G.A. (eds.) (2012). Robot ethics: The ethical and social implications of robotics, MIT Press, Cambridge, MA.

Mikhail, J. (2007). Universal moral grammar: Theory, evidence, and the future, Trends in Cognitive Sciences 11, 4: 143–152.

Morgan, E. (2011). The descent of woman, Souvenir Press, London.

Pereira, L.M. & Saptawijaya, A. (2016). Programming machine ethics, Springer, London.

Scanlon, T.M. (1998). What we owe to each other, Harvard University Press, Cambridge.

Shanahan, M. (2015). The technological singularity, The MIT Press, MA.

Tzafestas, S.G. (2016). An introduction to robophilosophy, River Publishers, New York.

Vardi, M. (2012). Artificial intelligence: Past and future, Communications of the ACM, 55, 1: 5.

Wallach, W. & Allen, C. (2008). Moral machines: Teaching robots rights from wrong, Oxford University Press, Oxford.

4 Robotization and medical ethics

The key ideas in this chapter

1. Tools that can change the genetic codes of all living organisms will change the process of evolution.
2. This new technology can be used to edit genetic code in the same way as we use text-editing tools to edit text.

Introduction

This chapter is about the relationship between ethics, artificial intelligence, intelligent robots and genome engineering. Throughout history, new medical technology has always generated ethical debate. There are many examples: blood transfusions, the use of antibiotics, the use of endoscopic instruments and so on. This time, the source of the ethical debate is genome engineering.

There is already a debate about robotic carers. How should we design such robots for application in the healthcare sector? Should we accept them without imposing reservations? Should we prevent the introduction of such robots into the healthcare sector? In this chapter, we take as our starting point Van Wynsberghe's (2015) perspective, which represents a middle way. According to this perspective, we should consider the development and design of these robots, while at the same time developing, and debating the implications of, ethical scenarios. We do this by considering a range of practical examples and theoretical points, as well as the practical benefits of these robots, in the context of an ethical discussion. Accordingly, this chapter does not consider robot technology in itself, but rather the ethical implications of that technology. The objective of the chapter is to generate new ethical knowledge that can be applied when robot technology is developed, designed and applied in the health sector. Accordingly, this chapter envisages that this knowledge could be used to develop robotic carers and medical robots that are based on values that have been developed in the health sector ever since the time of Hippocrates.[1]

Decision makers all around the world see robots as the solution to the impending shortage of people who (a) possess essential skills and (b) can provide them at a price that is affordable both for individuals and society as a whole (Nørskov, 2016: 99–121). Most OECD countries are facing the demographic problem of a growing older population and a shrinking group of the younger people needed to pay for welfare services. Anticipated longevity is increasing, and birth rates are declining (Doudna & Sternberg, 2018; Kass, 2018). These factors are exacerbating the challenges faced by many countries. The immediate problem is a lack of healthcare personnel. To solve this problem, many countries have looked to robotic carers and medical robots. Hospitals are already using robots for various functions such as rehabilitation, transporting drugs and food, lifting patients and as a means of communication in surgical operations that are conducted remotely (Sharkey & Sharkey, 2011).

Robotic carers have also been introduced in the elderly care sector. Interacting with robotic pets such as Sony's AIBO or the PARO robot baby seal has been shown to be psychologically beneficial for elderly people. Service robots such as Aethon's TUG and HelpMate are already used in American hospitals to transport drugs (Van Wynsberghe, 2015: 1).

To be clear, it is **not** our opinion that all healthcare roles either can or should be replaced by robots. We must find a balance that ensures that everyone can be cared for in a respectful way at a sustainable price. In addition, there are many ethical challenges associated with the robotization of the healthcare sector. We should also not ignore the fact that some patients may be treated disrespectfully and unethically by human carers (Borenstein & Pearson, 2011: 257). There are many reports of disrespectful treatment of elderly and unwell patients both in care homes and in their own homes (Tronto, 2010). It is important to bear this in mind when discussing the ethical challenges of robots in the social and healthcare services.

A good deal has been written about the ethical challenges of using robotic carers instead of human carers to care for elderly and/or sick people. One challenge is social isolation (Sparrow & Sparrow, 2006), while another is physical human contact (Coeckelbergh, 2010).

Many of the ethical challenges about the use of robots in the health sector seem to derive from a fundamental uncertainty about what we mean when we talk about "care". If robots take over many functions in the healthcare sector, we will be able to reflect better on what human care for sick and/or elderly people really means. If we use robots to lift and to move patients, will that have a negative impact on human interaction with these patients? Most people would probably answer no to this question, even though we would then need fewer warm hands to lift patients. Will it reduce the level of care if we use robots to communicate with patients who are elderly, sick or otherwise in need of care? Most people would probably answer yes to this question, because communication seems to be more closely related to the concept of caring than the physical act of lifting a person. As robots enter the sector, we can reflect on the caring values and functions that we want to keep for humans, and those that can be performed advantageously by robots. According to Turkle (2011), one challenge may be that we expect higher standards from the technology we apply in the healthcare sector than we expect from the humans responsible for providing care in the healthcare sector.

At a minimum, we can consider care from two perspectives: practical and theoretical. From a practical perspective, we are concerned about the quality provided in the processes and activities that constitute care in the healthcare sector. Such activities might include, for example, nursing, diagnosis, prescribing of medication, dispensing of medication, keeping logs, administration, cleaning, maintenance of technical equipment and so on (Sharkey & Sharkey, 2011: 267–282).

In the care sector, as a theoretical concept relating to ethical issues, we are concerned about treating the person in need of care with respect, about taking responsibility for others and about the person being cared for having a life worth living (Benhabib, 2004).

Both practical and theoretical access to robot ethics in the health sector will benefit robots in three areas (roles). First, robots will be a preferable solution where caring roles are boring and repetitive. Here both the patient and the carer will benefit from having robots taking over some, if not all, roles. Second, it will be advantageous at least for healthcare personnel if robots perform tasks that may be hazardous for the carer, e.g., cleaning a room using products containing potentially dangerous chemicals. Third, it will be advantageous for both parties – the carer and the person being cared for – if robots take over functions that may be damaging for all involved. One example could be lifting obese patients. Here both the patient and the carer are at risk of injury. The use of various kinds of lifting robots has the potential to improve both safety and the quality of care.

In the light of both practical and theoretical considerations applicable to the care sector, we can see that robots used in the healthcare sector can improve the quality of care for many people. In this chapter, we will refer to practical examples, theoretical points and practical benefits in relation to the ethical problems that may emerge when robots are introduced into the healthcare sector and how this may affect the quality of care.

The question we investigate in this chapter is: what ethical problems arise when artificial intelligence and genome engineering are applied in the health sector?

The objective of the chapter is to generate new knowledge that can be applied when robot technology is developed, designed and applied in the health sector.

We have illustrated the introduction in Figure 4.1, which also illustrates how we have structured this chapter.

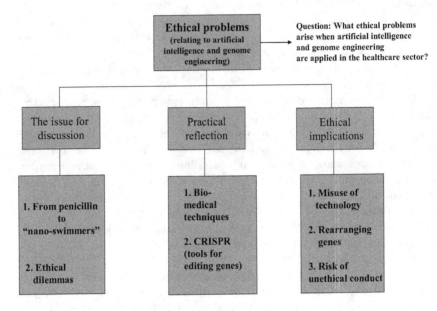

Figure 4.1 Robotization and medical ethics.

The issue for discussion

From penicillin to "nano-swimmers"

Penicillin was discovered by the Scottish microbiologist Alexander Fleming in 1929, although the work necessary to enable large-scale production did not start until 1938. During the Second World War, penicillin saved thousands of soldiers' lives. It seems reasonable to assume that nanotechnology and medical robots will have a similar effect on public health. In our own times, nanomedicine and nanocomputers offer hope for eradicating many of the diseases we fear today. In order to achieve this, it will be important for scientists in many fields – medicine, biology, technology, chemistry and so on – to work together towards this goal. The development of nanopumps, nanocomputers and nano-swimmers will depend on the integration of various fields of knowledge.

In the future, it is probable that instead of a surgeon or a medical robot performing a medical procedure, a "surgical nano-submarine" will be injected into the veins to perform the same procedure. This surgical submarine will then be guided by an internal "GPS system" and move towards the area where the intervention is to be performed. This may concern the administration of medication, the removal of a polyp, the removal of cancerous tissue, the cleansing of blood vessels, the treatment of heart valves, etc. One can also imagine the "submarine" remaining in the system to monitor and, if necessary, carry out a supplementary intervention, long after the initial intervention has been completed. According to Moore's Law, nanorobots will eventually be mass-produced at low cost (Moore's Law also explains how low cost PCs and other digital products have become available in recent years). When this situation occurs, each person will be able to have their own individualized "nanorobot", not necessarily for surgical interventions, but for monitoring hundreds of variables in the body, which vary within given parameters. This personal nanorobot can then be connected to medical centres that monitor when critical thresholds move above or below upper or lower limits. In such cases, the individual will be called in to their doctor for a closer examination. The use of such medical tools will most probably greatly improve health and increase longevity. The individual will also be able to control their blood pressure and blood sugar levels more effectively. However, the question here is what ethical challenges would be posed by such improvements in the general health of the population, as well as an increased life expectancy. We would end up with an even older population that would come in addition to the current trend of older populations.

The scenario described above shows that increased life expectancy by means of artificial intelligence, and by the use of various types of robots, is a realistic possibility in the not too distant future. Both Ray Kurzweil and Eric Drexler believe that nanocomputers will be able to operate at the cellular level and repair any damage that has occurred (More & Vita-More, 2013: 204–214); de Grey also believes that nanocomputers operating at the cellular level appear to be a feasible scenario (2013: 215–219). One must expect that what is technologically feasible and desired by many will eventually become a reality. However, there are some key ethical issues raised by this scenario. Will a life-prolonging technology necessarily be good for humanity? We can conclude at the least that this technology will be sought after by anyone suffering from a serious illness who is risking death if they do not receive treatment made possible by the technology. Should these people also have equal rights to medical treatment, just like everyone else? The question may seem to be hard-hearted, but with limited resources, it is a necessary question to ask. If a life-prolonging technology is used on an already ageing population, then it may result in resource scarcity, negatively affecting the younger part of the population. Leon Kass is perhaps the foremost advocate of promoting such a life-prolonging technology (Kass, 2018), while the foremost voice against this technology is perhaps Aubrey de Grey (2013: 215–219).

Logically and rationally, it is easy to argue for a technology that extends life to a hitherto unknown length, from the perspective of the person who wants to live for as long as the new technology makes possible. On the other hand, many will instinctively feel sceptical towards the idea that people may be able to live until they are 150–190 years old. Many will probably think that it doesn't feel right to technologically extend one's life to such an age. In this context, one can cite de Grey (2013: 216), where he in the following quote actually supports Leon Kass: "—ethics is ultimately more about instinct than about logic". If this ethical perspective were to be taken for granted, however, one would have to disregard ethics as a field of knowledge based on epistemology, axiology, ontology and methodology. Grey's view on ethics seems to suggest that ethics depends on each person's individual perception. Basing ethics on instinct is interesting, because what is instinct? Gregory Bateson (1972: 38–58) discussed this in his book *Steps to an Ecology of Mind*. Bateson states, and we agree with Bateson: Instinct is an explanatory principle, but an explanatory principle cannot explain another principle such as ethics being instinctive. It is, as Newton expresses, "hypothesis non fingo"[2] (Bateson, 1972: 38). Should doctors, nurses and others working in the health services only follow their own instincts and not base their ethical actions on established ethical knowledge? Indeed, if we were to agree with Grey's and Kass' view, this would be the result. If such thinking were allowed to spread across the health sector, one could then say that medical ethics were being left to the individual health worker's personal "instinct". It is obvious to anyone that such a situation would lead to complete chaos. One cannot rely on instinct as an explanatory principle of medical ethics, just as one cannot leave any legal judgement to the discretion of an individual. Ethics is just as much a field of knowledge as are technology, medicine, health and care, and other fields.

The question that emerges from what is described above is the following: if we can develop a technology that can extend life, physically and mentally, far beyond "normal" life expectancy, should one even apply such a technology at all? If it is used, how long should a person live? Is such a medical technology really a Pandora's Box[3] that should remain closed? On the other hand, who are we to even ask such questions about how long someone should live, that is, deciding that someone should die before they absolutely needed to die. Asking questions is not a neutral activity; this is something of which Socrates was well aware when he asked questions that aroused the ire of the leading men of Athens, a step culminated in him having to drink a goblet of poison resulting in his death.

Darwin's theory of evolution by natural selection may be strongly challenged by the new life-prolonging technology. New discoveries in nano- and biotechnology may completely change ideas concerning evolution as a natural selection process. We are close to the point in medical history where conscious technological choices may replace the idea of evolution as being governed by natural selection. When medical technology can diagnose genes and then consciously intervene to control genetic development, then the Darwinist idea of evolution that presupposes natural selection will have become an anachronism, and advances in medical technology will strongly affect the processes of evolution.

Genetic engineering will not only result in prolonged life but also reduce and partially eradicate many diseases, so that people will be able to greatly improve their health. Of course, such technology can also lead to greater differences between those countries that can afford this medical technology and those that cannot. It can also lead to major differences within countries, where those who can afford such life-prolonging technology will improve their health and live longer than those who cannot afford it.

In addition to the aforementioned medical technology, we will also witness new technology that is able to produce spare body parts such as organs and limbs. Genetic engineering and new technology that can produce body parts using, for example, 3D printers, will also result in new perspectives on medical ethics, especially within bio-ethics and bio-politics. Bio-ethics questions many aspects of this development, and bio-politics concerns, amongst other things, the political aspects of medical technology and medical ethics. Bio-politics addresses issues of the type: if it is possible to develop such a medical technology, do we want to use it? The use, side effects and consequences of such technology are overwhelming and impossible to predict. Darwinist evolutionary development took a long time, indeed a very long time, while the evolution made possible by new medical technology will greatly accelerate the evolutionary process. Can aesthetics and beauty become part of the moral argument in genetic engineering? If aesthetics become part of the ethical agenda, then morality will have collapsed, and will have veered away from the question of what is right and wrong, towards what is beautiful and ugly.

If we develop a society where those with financial resources can buy both beauty and a longer healthy life, while others who are poorer do not have this possibility, can society endure such a development? Can we, in such an imaginary society, envision tension and a potential conflict between the "beautiful" people with good health and the less affluent with poor health prospects? When money not only provides the gateway to a good life but to life itself, then by means of such a health technology will we have created a monster that will result in a new underclass and a new upper class, based on aesthetic and health variables?

It is no longer a question of whether we will end up with such a medical technology as is described above, but *when* it will become a reality. Another question is: what can we know about the consequences of such a technology? The next question is, if possible, even more important: what do we not know about the consequences of this technology? What do we have to know to allow the development of such a technology? There are some things we can know, but there are probably even more we can't know, and some things where we don't even know what we don't know what to ask. The point is that the emergence of this new medical technology will result in major challenges, not least in the discussion of medical ethics that will come in the wake of this technology. The question we have to deal with, even though we have no answer to it, is the following: what is the right thing to do when we do not know the consequences of what we are doing? Not doing anything is of course one possibility, but not a realistic scenario, for many reasons. The main reason is that technological, economic and political forces will push for the development of new technological opportunities. In addition, if people can improve their lives using this technology, they will also push for further development of the technology. Therefore, not doing anything is not a possibility. BUT: what should one do?

Ethical dilemma

The problematization of the issue of new medical technology and related ethical questions and dilemmas tends to divide theorists and writers into two groups. First, there are those who believe that the new medical technology can be a breakthrough in medical research; and second, those who are sceptical towards the new technology and believe it can lead humanity into a serious crisis. However, much of the technology already exists that can revolutionize medicine and the health and care sector. Thus, those sceptical of new developments cannot "smash the machines", in the manner of a modern Luddite movement,[4] or a neo-Luddite movement,[5] as this is not an option. The question is perhaps whether there is an alternative to letting the new technology develop freely? There is a middle way which will involve the application of medical ethics. This would involve new medical technological developments always being monitored by ethical committees, which would submit reports, possibly giving their approval, before further development and application of the new technology. Similar systems have already been established in many countries, at the national level, but not to the same extent in the individual enterprises that develop this technology.

Regardless of whether or not ethical committees are established, one will still face medical technological changes; some will involve incremental changes, while others will be radical innovations. The incremental changes and radical innovations will greatly affect both individuals and society. The greatest dilemma will perhaps concern the rights of the individual and their freedom to choose treatment, if they have the resources to pay for it, and how such freedom of choice and use of treatments serve society as a whole. For example, should genetic engineering during the prenatal stage be allowed?

When singularity occurs in the future and new technology becomes available (Johannessen, 2021), the ethical challenges may potentially be greater than the technological challenges. The development and implementation of new medical technology will lead to a politicization of medical ethics related to the new technology. The main question will probably be: to what extent is that which is technologically possible also ethically justifiable? If it were possible to develop new medical technology, which will be able to help one or more people, would it be ethical to prevent the development and use of such medical technology?

Practical reflections

Biological medical techniques

Intelligent robots are not the most challenging issue facing future medical technology and, subsequent to that, medical ethics. The greatest challenge is related to biological medical techniques that will put a rocket up evolutionary process.

Using biological technology, it will be possible to eradicate diseases such as cancer, heart disease and malaria, say Doudna and Sternberg (2018: 117–154). We can only imagine what new discoveries will emerge from this new technology, when the fields of biology, chemistry, technology and medicine interact with each other.

Envisaging future scenarios relating to this field will, in part, prepare us for future developments. In our view, the ability to envisage such scenarios is the cohesive force that connects the present to a potential future, because it prepares us for the problems and challenges we will face when the new technology is developed and implemented. The point of this brief description is to point out that in many fields, science and policy gain insights from the envisioning of future scenarios. In this section, we will discuss some possible future ethical issues related to the new technology that will be created by the interaction between biology, chemistry, technology and medicine.

Biotechnological tools[6] now make it possible to modify DNA inside cells (Doudna & Sternberg, 2018: xv) and modify the genetic code of all living organisms. It is now possible to modify not only the genetic code and DNA structure of a single human being, but of all humanity, as easily as editing a text using text-editing tools. Using gene-editing tools, it is possible to change, edit and delete genes in all living organisms such as humans, dogs, apple trees, fish, pigs and so on. It is highly probable that this technology will affect human evolutionary development. This scientific breakthrough and creation of a practical tool enable specialists to undertake genetic engineering and take control of the evolutionary process. One might say with the emergence of this new technology that Darwin's evolutionary theory has now been stowed away in the historical archives of science, and biologists, chemists and genetic engineers have taken over the evolutionary process using gene-editing tools such as CRISPR. However, this is not so much an imagined future scenario we are describing here, but an emerging reality (Doudna & Sternberg, 2018). By modifying a gene that controls muscle formation, it is now possible to create animals with powerful muscular frames, or cattle that yield more meat. In other words, it is now possible to change the genetic code of any animal we use for our food, to improve quality and increase the quantity of food yielded; another area where agricultural production could be increased is through the genetic modification of fish on fish farms. Many foods have already been genetically modified such as cereals that are more tolerant of temperature changes and disease-resistant rice, which is vital to the diet of millions of people. This is done by specialists modifying the DNA structure of plants using the new genetic engineering tool, CRISPR. Thus, it is only our imagination that can limit the scope of opportunities presented by this new technology. Currently, genetic engineering tools are used to modify the DNA of plants and animals, but in the near future gene-editing of humans will be possible (Doudna & Sternberg, 2018). It can be imagined that this new tool may be used to modify those genes that influence logical intelligence, so that it will be possible to create humans with super intelligence. However, in this context, the ethical question is: do we wish to see such a development? Will it be good or bad for mankind to put a rocket up evolutionary process in this way?

CRISPR – A tool for editing genes

In the future, scientists are confident they will be able to eradicate mosquito-borne diseases such as malaria by using CRISPR, something that will be welcomed by those at risk of such diseases. So far, gene-editing technology has been used to cure cystic fibrosis and some forms of blindness. It is hoped in the future that this medical tool will soon be able to cure other diseases as well such as HIV/AIDS (Doudna & Sternberg, 2018: xvi–xvii). Surely, eradicating such diseases must represent progress for medical research and practice? Using CRISPR-based therapies will make it possible to correct mutations responsible for a number of diseases, thus preventing illness and death in many cases.

It seems logical and expedient to use such a therapeutic tool that can relieve people's suffering, and save people from deadly diseases. This will soon be possible in the future using genetic-editing technology such as CRISPR. It will also be possible to use such technology in the early stages before a disease develops, to edit and repair mutated genes. In some cases, it will also be possible to carry out medical procedures on human embryos. For instance, the gene structure of the unborn child could be modified to prevent or reduce the possibility of cardiovascular diseases, cancer, Alzheimer's, diabetes, etc. However, will such interventions raise ethical issues? One issue that immediately springs to mind is the quest for perfection and the rejection of unborn children that are not "normal". The ultimate consequence of such thinking may lead to the creation of a "pure race". This quest for perfection would also draw attention to those in society who are not so "perfect". This could, in turn, lead to a situation where those people with congenital, physical or mental disabilities would be viewed as having less value.

For hundreds of thousands of years, evolution has been governed by two forces: natural selection and random mutation. By intervening in this evolutionary process, evolution processes will include a third element. We now have a medical tool that can edit the DNA structure of any living human being and also that of future generations. In the future, will the power inherent in such a tool be exploited by some people to create, for example, super-soldiers, super-intelligent scientists, or, for example, edit the DNA structure of humans so they become less violent and criminal? The new technology is already available to proceed down this road. However, do we want to fully exploit this new medical technology? Should we use the technology before we know all the consequences which may result from it? We can say with certainty that this is such a complex issue that it is impossible to comprehend all the possible future consequences and side effects. Should we stop the development of the technology? What ethical consequences can we reveal at the present stage of development of this medical technology?

Ethical consequences of using medical-genetic tools

We will investigate here three areas in relation to the use of medical-genetic tools: abuse of the technology, restructuring of genes and the risk of ethical derailment.

Abuse of the technology

If genetic-editing technology is as easy for specialists to use as a text-editing tool is to use for lay people, is it probable that such technology will be used in many different kinds of situations? The answer really lies implicit in the question, so let's rephrase it: under what conditions will genetic-editing technology not be abused to serve someone's interests? On the one hand, the technology could be banned, because it has the potential to be abused. On the other hand, the opportunities for advances in the field of medicine are so outstanding that it is difficult to imagine the whole technology being banned. When a technology is efficient and easy to use, the probability that it will be used is great (Kurzweil, 2005). Yet, it is not difficult to imagine the nightmarish scenarios that could emerge from unchecked use of gene-editing tools. Nations would be able to design their own "soldiers-of-the-future", industries could produce compliant and deferential workers; moreover, it may become possible to use the technology to "edit out" aggressive tendencies, criminal behaviour and other so-called "abnormal" behaviour. The idea that dictators who advocate antihuman race ideologies will have the opportunity to apply this technology on a large scale makes the revelations of the apocalypse seem like a slight calamity in comparison; for instance, how would a modern Himmler use this technology? When the nations of the world seem unable to prevent the proliferation of nuclear weapons to new states, it is difficult to foresee how it will be possible to prevent the proliferation and abuse of a simple technology that can change the genome structure of the human race, literally creating a new type of human.

On the one hand, genetic technology will be able to help cure diseases. On the other hand, one can envisage thousands of little "Himmlers" experimenting with human genetic material. Are the benefits of such technology, which are obvious, worth the risk of abuse? Is the new technology a monster in disguise, or is it the saviour of mankind, where everyone can live a life without serious illness? One can only imagine what some state leaders might do with such technology. It seems unlikely that they would be transparent regarding any experiments they might carry out. The assumption is that various state leaders, not only in dictatorships, will keep the experimentation and application of the technology secret. We would not know anything until it is too late, once the results and consequences appear.

Restructuring of genes

It is only when a technology has demonstrated its success that one reflects on its limitations and the ethical issues in which the use of the technology may result (Parrington, 2016). At this point in time, it will almost certainly be too late, because technological success will promote the demand for the technology (Kurzweil, 2005). If the technology is also simple, easy to use and has many applications, it will be almost impossible to stop. The atomic bomb and atomic energy are just one of many such examples. It is against this background that medical technology, in general, and genetic diagnoses and restructuring of genetic material, in particular, should be discussed, before it is developed into a technology that can be easily used and spread.

The ethical concerns and challenges of restructuring genetic material are problematic. Despite these challenges, it is reasonable to assume that the demand from individuals and countries will be very great, because genetic restructuring can be of great help to people suffering from hereditary diseases, as well contributing to increased agriculture and fisheries' production (Carey, 2019).

The following questions raise some of the ethical dilemmas we are facing: should parents be allowed to choose the genetic traits of their unborn child? Should parents be given the option of a genetic diagnosis of their unborn child? If a genetic mutation is discovered in the embryo that may result in a serious disease later in life, should genome modification be allowed?

Genetic restructuring will not only be used to create genetically modified crops that can withstand climate changes, but possibly also to genetically modify humans for various reasons. Genetic engineering in agriculture is already allowed in many countries around the world (Parrington, 2016). To date, there are to my knowledge no countries that have allowed human genetic engineering; nonetheless, if policy-makers in some countries find it to their advantage to proceed down this road, then it is probable we will see this development. However, this road is strewn with moral hazards and legal complications, where ignorance prevails. Therefore, we will be on shaky ground, both morally and legally speaking. If the technological risks are too great and possible societal challenges formidable, will the benefits outweigh the disadvantages in the development and use of this type of technology? Will we be able to untangle ourselves from the ethical and legal dilemmas posed by the future development of genetic engineering and the restructuring of genetic material? For instance, what will happen if genetically engineered bacteria goes astray? Could this trigger an epidemic killing millions of people? On the other hand, one can imagine that gene-editing could eventually eradicate diseases such as cancer and malaria.

Can and should one stop a technology that can eradicate several of the major diseases in the world? Possibly, one way of problematizing the issues involved regarding the editing and restructuring of genetic material may be to carry out financial impact assessments of this technology. It seems reasonable to assume that the technology cannot be stopped, because it has very high cost-benefits in many industries and for many people. On the other hand, one can come to grips with the consequences of this technology by investing in ongoing, ethical, legal and technological impact studies.

Some concrete suggestions for getting to grips with the problems posed by the technology and be in advance of developments could be:

a. To establish precautionary regulations in relation to ethical and other guidelines, and also determine the financial responsibility of the researchers and entrepreneurs who implement the technology.

b. National health administrations can establish ethics committees with a specific focus on the technology. They can then provide ongoing input to the public authorities about the possible consequences of using the technology.

c. A legal examination of the consequences of using the technology, implemented by ministries of justice.

d. International forums, such as the UN and WHO, can establish permanent committees focusing on the consequences of using the technology.

e. In the same way as nuclear disarmament and non-proliferation agreements have been implemented, the UN could lead a committee to prevent the spread and abuse of the technology.

f. Universities could introduce mandatory courses on this technology and the possible consequences that the technology may have for humanity, both positive and negative.

Risk of ethical derailment

In the future, embryo diagnosis may quickly lead to rejecting embryos that do not meet a desired "standard". On the other hand, it can prevent the child from developing genetically dependent diseases. Consequently, a balance must be struck between reducing hereditary diseases on the one hand, and rejecting human embryos because they do not meet some "standard" on the other hand. Who would be responsible for setting such a "standard" regarding what is desired or not desired? In some countries, baby boys are preferred to baby girls. Should parents be allowed to choose the gender of their unborn child?

No one will dispute that looking after children with serious genetic diseases can be a great burden for many parents, but who should decide exactly what genes a child should have? It is not a big step from being able to choose the gender of the unborn child to choosing traits such as physical appearance, behavioural traits, or level of logical intelligence. If birth clinics acquire gene-editing technology, what will prevent them from providing services that parents desire? Of course, such tests or diagnoses may be viewed as unethical and be banned by law. However, if the technology exists, and there is demand for the technology, then even when there is legislation restricting its use, it will be difficult to fully control developments.

Perhaps the greatest risk posed by the new technology, which can determine the lives of many people, is the fact that biologists, bio-engineers, biotechnologists, biomedical engineers, etc. have neither the experience nor the expertise to examine and assess the ethical and philosophical issues related to this area. Jennifer Doudna, a biochemical researcher, and one of those involved in the development of the new bioengineering tools used around the world in genetic diagnosis and editing, says: "I was not used to asking myself these questions in my day-to-day life as a professor and biochemical researcher" (Doudna & Sternberg, 2018: 197). It is perhaps a little late to reflect on the consequences of using nuclear weapons after they have been used. Likewise, it's a little late to reflect on the consequences of using CRISPR two years after the tool was developed, which is Doudna says she did (Doudna & Sternberg, 2018: 197). It is understandable that talented researchers mainly wish to communicate with other researchers in their own field of study, paying little attention to ethical and philosophical questions of their research, but is it acceptable from a social perspective?

Should a principle of equality apply to the use of this technology, i.e., should everyone have access to the technology, regardless of their financial resources? Editing of genetic material will appeal to many parents, who perhaps look back on their own lives and hope their children will not have to face the same uphill struggle through life as they had to. Are we in the process of developing a technology that will soften the trials of life? Are we in the process of creating the Teflon man, the shining prince or princess in their own little castle? Undoubtedly, if given the opportunity, many parents will want to use gene-editing technology to modify the genes of their unborn child. In this context, possibly aesthetics will be given priority over ethics? Will it be possible to hinder the development and use of this technology by law? Will parents who have the resources be able to travel to other countries that do not have such strict laws concerning human genetic engineering? We have seen something similar where there are different abortion laws and euthanasia laws in different countries. Why should this technology be any different? If this situation evolves, will the rich be able to carry out genetic editing of their unborn child to produce their idea of the "perfect" child, while those who are less well-off will have to rely on the random aspects of evolutionary processes? Will we eventually see a distinction between the perfect children of the rich and those children of the less well-off, if such technology and legislation are established in some countries, but not in others?

Once this biological innovation is accessible to entrepreneurs and investors, it will be very difficult to close the door on the technology, and go back to studying the ethical and philosophical implications of this bio-economy. We have cases from China, where research groups have used the CRISPR technology in an attempt to edit and improve the beta-globin gene, which is part of the haemoglobin protein that carries oxygen around the body (Doudna & Sternberg, 2018: 214). A similar technology could be used to improve the performance of athletes, while those athletes who had not undergone such gene-editing would be at a disadvantage. On the other hand, there are people who suffer from a disease where the blood transports too little oxygen, and in such cases, the experiment could be of great help. However, the point here is that once the technology has been used, it will be difficult to reverse it, or only use it in the areas for which it was intended. It should be noted here though that the above experiment carried out in China in 2015 was not wholly successful. However, this is not the point here; the point is that experiments are being carried out on human embryos, and the consequences of these experiments can be many. In China, the result was that further development of the technology was needed, if they were to attain the desired results. However, the risk here is not so much that the experiments fail, but rather that sooner or later they will be successful, because such a technology may ultimately transform what we believe constitutes a human being. The link between bio-technology and artificial intelligence can create a whole new human being, which we cannot yet imagine. We have most probably passed the threshold where we can turn and go back. It may be only a small step from the laboratory experiments on human embryos in China aimed at curing disease, to medical applications.

One can ask what the experiments within biotechnological research are that we never get to hear about. While regulatory authorities in the United States and Europe have ensured that gene-editing in human embryos will be subject to strict controls, it is imaginable that entrepreneurship and investment will drive innovation processes forward to meet the demand for products within this sector.

Governmental authorities may build a dam to stop the flow of these developments, but the dam will most likely be cracked by the force of profit. Entrepreneurs will crawl through the cracks and start up medical enterprises that will probably be flooded with investment, given the likelihood of high profits. This metaphor illustrates the fact that when opportunities for high profits exist, then capital will flood into the sector.

The rate of spread of this type of knowledge in the global economy is very high. If research groups elsewhere, for example in China, carry out new experiments that some investors find profitable to invest in, then the knowledge that emerges from these research groups will quickly spread to other parts of the world; it's just a matter of opportunity and financial resources. There are rumours of similar laboratory experiments to the one mentioned above being carried out in China, say Doudna and Sternberg (2018: 217). English researchers have also applied for a licence to edit genes in human embryos.[7]

The road to hell is paved with good intentions, says the proverb, which may aptly be applied to gene-editing tools such as CRISPR. The researchers who developed the tool obviously had good intentions to improve people's lives. However, the possibility of the spread of this technology is high, which may result in the tool ending up in the wrong hands, i.e., those of people without good intentions. Going down the legal path is of course one way of limiting the spread of this technology, but this will definitely not stop the spread, because so many financial resources are involved in the various uses of the technology.

The risks involved in using gene-editing tools can be summarized as follows:

1. The risk of the spread of the technology: In the near future, gene-editing will be used in the medical industry and medical clinics (Doudna & Sternberg, 2018: 222–223).
2. Mutation risk: Although the cells in the body undergo endless mutations through a life cycle, controlled genetic editing can result in inflicting further unforeseen mutations on the cells with serious consequences (Doudna & Sternberg, 2018: 225).
3. Knowledge aversion: It seems to be an impossible aim to gain complete knowledge of the consequences of genetic editing. The argument against the need to gain full knowledge of the consequences is that much medical research and treatment has been carried out earlier with good results, before we have known the full consequences. Therefore, there seems to be a growing interest in continuing genetic-editing and spreading the technology, without having a complete knowledge of the consequences.
4. If we can then we should: Technology is becoming increasingly fine-tuned and integrated with artificial intelligence and intelligent robots. It therefore appears that genetic-editing can be developed into a mass industry. This applies not only to the human genome mass but also to plants, fish and animals. One argument that is put forward is that the problem of starvation can finally be solved. However, other developments may also easily lead to the quest for the "perfect" human being, whatever this may be.
5. Genetic diseases: The strongest argument for the use and dissemination of the technology is that it can prevent the genetic mutation that leads to diseases and that it can enable the modifying of the genes of unborn children to cure genetic diseases. However, once the technology has been used for good purposes, the spread of the technology may also lead to it being abused for other purposes.
6. What is natural and what is unnatural: Over approximately 3 billion years, the human genome has evolved from a bacterial stage to where we are today. During this time, millions of mutations have occurred. These mutations, which have also led to serious diseases, are considered part of the natural selection process, while the more rational gene-editing, which can save many victims of so-called natural mutation, is considered "unnatural". Support for the use of gene-editing technology seems to be gaining support in many environments, if we disregard the Christian cultural sphere (Doudna & Sternberg, 2018: 27–228). The support for this technology will promote the development and dissemination of gene-editing technology, before we know the full consequences of the use of the technology.
7. Robots and gene-editing: There is no consensus on what constitutes a robot. In the near future, it may be possible for gene-editing tools to be implemented in intelligent robots, which can then perform gene-editing as desired. Assumption: Given the evolution of knowledge in gene-editing,

the possibility of global distribution and the low cost of technology applications, this technology will lead to individual, national and global security challenges. There are many examples in the past where technological development and innovations have failed and led to greater uncertainty for individuals, organizations and countries. The environmental issue is perhaps the biggest challenge today, and it shows that not all new technologies can necessarily be equated with progress. Global warming and air pollution in the big cities are two concrete examples that demonstrate that new technology is not necessarily always positive. Today, risks are associated with the use of nanorobots as well as with gene-editing, both from the standpoint of technical security issues, but also in relation to ethical issues (Allhoff et al., 2010; Doudna & Sternberg, 2018).

8. Software and design: If intelligent robots take over gene-editing tools, then developments within gene-editing will be left to software engineers and hardware design engineers. The risk is that the ethical perspective will fade into the background and technical feasibility and profits will receive all the focus. Millions of codes written by teams of programmers, no matter how skilled they are, have the potential for technical errors. The focus of these engineers will be on technical quality, with the risk that ethical issues will be given less attention.

Conclusion

This chapter has explored the following question: what ethical problems arise when artificial intelligence and genome engineering are applied in the healthcare sector? The general answer to this question is that ethical problems of all possible types will arise, concerning everything from the risks associated with the use of artificial intelligence and genome engineering, to existential questions about the nature of life itself. We have chosen to respond in more detail to the research question, by dividing the answer into three parts: practical implications, theoretical implications and practical benefits.

Practical implications

At a more practical level, three types of ethical problems arise.

First, we can view the ethical problems linked to artificial intelligence and genome engineering in light of the professional ethics governing the various professions in the healthcare sector.

At a practical level, these professional ethics can be linked to three areas: respect, responsibility and dignity. In this way, patient care will be optimized.

Second, the ethical problems can be linked to the moral codes and rules that are implemented in intelligent robots of various kinds, from nanorobots to the use of artificial intelligence in genome engineering. What happens in this type of problem is that robots are obedient instruments for the coded orders that are designed into their software.

Third, we can say that ethical problems will concern robots' abilities to be ethically self-aware, i.e., raising the question of to what extent intelligent robots will be able to adjust their ethical rules to cope with new or unanticipated situations.

Theoretical implications

Proposition 1: The deeper we manage to dig into the ethical problems relating to medical ethics and robotization, the more likely it is that these tools will be safe for patients.

Proposition 2: The more we are prepared to tackle the ethical problems that are created by new technology, the better we will be able to deal with the change processes we will face in the future.

Proposition 3: If we produce intelligent robots whose logical and rational intelligence surpasses our own, then the ethical problems that we consider insoluble will be transformed into ethical problems capable of solution.

Proposition 4: As robots become more intelligent, so the solutions to ethical problems will become simpler, because we will have the assistance of the intelligence that was the original source of the ethical problems.

Proposition 5: The future can be predicted if we develop/create intelligent tools that surpass human logical and rational intelligence by a factor of N.

Practical benefits

The greater the increase in the practical benefits experienced by users, the greater the likelihood that the development of this kind of medical technology will accelerate.[8] These practical benefits will be the feedback that will lead directly to an exponential growth rate in the development of medical technology, robots, artificial intelligence and genome engineering. This practical benefit will probably also outweigh ethical concerns about the use of this technology. Accordingly, it is important that we establish a robust ethical system before we reach this point of no return. Such a system will need to be capable of surviving one of the consequences of the law of accelerating returns, i.e., the idea that the use and development of technology increase as users experience increased practical benefits.

Notes

1 https://en.wikipedia.org/wiki/Hippocrates.
2 "Hypothesis non fingo" means to make no attempt at a hypothesis.
3 https://en.wikipedia.org/wiki/Pandora%27s_box.
4 https://en.wikipedia.org/wiki/Luddite.
5 https://en.wikipedia.org/wiki/Neo-Luddism.
6 Genetic engineering tools such as CRISPR-Cas9 (CRISPR for short) (Doudna & Sternberg, 2018: xv).

7 *Nature News*, September 18, 2015.
8 This is directly related to Kurzweil's "Law of accelerating returns" (Kurtzweil, 2005: 35–111).

Bibliography

Allhoff, F.; Lin, P. & Moore, D. (2010). What is nanotechnology and why does it matter? Wiley-Blackwell, Hoboken, NJ.

Bateson, G. (1972). Steps to an ecology of mind, Intex Books, New York.

Benhabib, S. (2004). The rights of others: Aliens, residents, and citizens, Cambridge University Press, Cambridge.

Borenstein, J. & Pearson, Y. (2011). Robot caregivers: Ethical issues across the human life. In Lin, P.; Abney, K. & Bekey, G. (eds.). Robot ethics: The ethical and social implications of robotics, MIT Press, Cambridge, MA. pp. 251–266.

Carey, N. (2019). Gene editing, Icon Books, New York.

Coeckelbergh, M. (2010). Health care, capabilities, and AI assistive technologies, Ethical Theory and Moral Practice, 13, 2: 181–190.

De Grey, A. (2013). The Curate's egg of anti-aging bioethics, in More M. & Vita-More N., (eds.). The transhumanist reader, Wiley-Blackwell, New York. pp. 214–219.

Doudna, J. & Sternberg, A. (2018). A crack in creation: The new power to control evolution, Vintage, New York.

Johannessen, J-A. (2021). Singularity innovation, Routledge, London.

Kass, L. (2018). Leading a worthy life: Finding meaning in modern times, Encounter Books, New York.

Kurzweil, R. (2005). The singularity is near, Duckworth, London.

More, M. & Vita-More, N. (eds.) (2013). The transhumanist reader, Wiley-Blackwell, New York.

Nørskov, M. (2016). Technological dangers and the potential of human-robot interaction: A philosophical investigation of fundamental epistemological mechanisms of discrimination, in Nørskov, M. (Ed.). Social Robots, Ashgate, London. pp. 99–121.

Parrington, J. (2016). Redesigning life: How genome editing will transform the world, OUP, Oxford.

Sharkey, N. & Sharkey, A. (2011). The rights and wrongs of robot care. In Lin, P.; Abney, K. & Bekey, G. (eds.). Robot ethics: The ethical and social implications of robotics, MIT Press, Cambridge, MA. pp. 267–282.

Sparrow, R. & Sparrow, L. (2006). In the hands of machines? The future of aged care. Minds and Machines, 16, 2: 141–161.

Tronto, J. (2010). Creating caring institutions: Politics, plurality and purpose. Ethics and Social Welfare, 4, 2: 158–171.

Turkle, S. (2011). Alone together: Why we expect more from technology and less from each other, Basic books, New York.

Van Wynsberghe, A. (2015). Healthcare robots: Ethics, design and implementation, Ashgate, London.

Appendix 1: Chapter on concepts

Asplund's motivation theory.[1] In brief, this theory can be described in the following way: *people are motivated by social responses* (Asplund, 2010: 221–229). The following statement may be said to be a central point made by Asplund's theory: *when people receive social responses, their level of activity increases.*

Asplund's motivation theory is consistent with North's action theory (ref. North's action theory). Understood in this way, it seems reasonable to connect the two theories in the statement: *people are motivated by the social responses rewarded by the institutional framework.*

Availability cascades. This refers to the idea that we are all controlled by the image of reality created by the media, because this image is easy to retrieve from memory.

Availability proposition. This may be expressed as follows: The more easily information enters into our consciousness, the greater the likelihood that we will have confidence in that information. In other words, we believe more in the type of information that is available in memory than the information that is not so readily available.

Behavioural perspective. This perspective focuses on the behaviour of employees as an explanation for the relationship between business strategy and the results obtained.

Boudon-Coleman diagram. This research methodology was developed by Mario Bunge (Bunge, 1978: 76–79) based on insights made by the sociologists Boudon and Coleman. The purpose of the diagram is to show the relationship between the various levels such as the macro- and micro-levels. For instance, it is shown how changes at the macro-level, such as technological innovations in feudal society, can lead to increased income at the micro-level. However, it was shown that technological innovations could lead to weakening of the semi-feudal structures because dependency on landowners was reduced. Consequently, the landowners opposed such changes especially in the case of technological innovations, which Boudon has shown in his research (Boudon, 1981: 100). Coleman (Coleman, 1990: 7–12) started at the macro-level, went to the individual level to find explanations and finally ended up at the macro-level again.

An important purpose of Bunge's Boudon–Coleman diagram is to identify social mechanisms that maintain or change the phenomenon or problem under investigation (as mentioned above, in Boudon's analysis of semi-feudal society). Bunge's Boudon–Coleman diagram may be said to represent a "mixed strategy"; Bunge says the following: *when studying systems of any kind (a) reduce them to their components (at some level) and the interaction among these, as well as among them and environmental items, but acknowledge and explain emergence* (see the chapter on concepts) *whenever it occurs, and (b) approach systems from all pertinent sides and on all relevant levels, integrating theories or even research fields whenever unidisciplinarity proves to be insufficient* (Bunge, 1998: 78). The purpose of this research strategy is to arrive at a deeper and more complete explanation of a system's behaviour.

Capabilities. Capabilities are for an organization what abilities are for an individual.

An organizational capability may thus be defined as an organization's ability to perform a task, activity or process. Operational capabilities enable an organization to make money in the here and now (Winter, 2003: 991–995). Dynamic capabilities, as opposed to operational capabilities, are linked to processes of change. Change and innovation are at the centre of dynamic capabilities.

Simplified, one may say that organizational capabilities are something an organization does well compared to its competitors (Ulrich and Brockbank, 2005). These capabilities are intangible and therefore difficult for competitors to imitate (Wernerfelt, 1984).

Cohesive energy. In a social system cohesive energy is "the glue" that binds the system together. Cohesive energy is the social mechanisms that make the system durable. According to systemic thinking it is the relationships and actions that bind social systems together. The rationale is that relationships and the systems of relationships may be said to control human behaviour. Social systems are held together (in systemic thinking) by dynamic social relations (e.g., feelings, perceptions, norms) and social actions (e.g., cooperation, solidarity, conflict and communication).

Co-creation. Co-creation involves working together to promote knowledge processes and innovation. If knowledge processes and innovation are essential for value creation in the knowledge society, co-creation is an important social mechanism for initiating, maintaining and strengthening these processes. The balance between competition and cooperation, embodied in the concept of co-creation, leads to constructive criticism and the necessary scope of knowledge that exists in the network so as to promote creativity and the innovative. Instead of a zero-sum situation, a positive-sum situation will be developed where everyone wins.

Collective blindness. Collective blindness may be said to be a form of collective arrogance, which results in irrational actions. Minor events slip under the radar, causing the system to not be fully aware of what is happening. Politicians' explanation why voters in a referendum vote contrary to what most of the power elite and the media advocated is an example of collective blindness.

Competence. Competence refers to knowledge, skills and attitudes.

Core competence. The concept was popular in the strategy literature of the 1990s. Core competence may be defined as: "a bundle of skills and technologies that enable a company to provide a particular benefit to customers" (Hamel & Prahalad, 1996: 219). More recently, core competence as a concept has been given less attention in the research on dynamic capabilities, and now there is more focus on the concept of *fitness*. The term *evolutionary fitness* is also used in the research literature in connection with technology, quality, cost development, market development, innovation and competitive positioning (Helfat, et al., 2007: 7).

Discontinuous innovations. These are innovations that change the premises of technology, markets, our mindset and so on. We know that sooner or later discontinuous innovations will emerge in the future (Hewing, 2013).

Dynamic capabilities. Dynamic capabilities stem from the resource-based perspective and evolutionary thinking in strategy literature (Teece, 2013: 3–65; 82–113; Nelson and Winter, 1982). The dynamic perspective attempts to explain what promotes an organization's competitive position over time through innovation and growth (Teece, 2013: x).

The original thinking concerning dynamic capabilities may be related to Teece et al. (1997). These authors defined dynamic capabilities as *an organization's ability to create, develop and modify its internal and external expertise in order to address changes in the external world.*

Dynamic capabilities are now seen as all the organizational processes, not only internal and external expertise, that contribute to an organization's capacity to adapt to change while creating the organization's future.

Explicit knowledge. This is knowledge that can be digitized and communicated to others as information.

Evidence. This may be results, such as research results, that can be relied on. However, it is also important to be aware of the fact that other evidence may be available without having to refer to figures and quantities, such as evidence that emerges from observations and good judgement without the assessment being quantified. Evidence-based research is research results that are based on approved and accepted scientific research methods.

Emergent. An emergent occurs if something new turns up on one level that has not previously existed on the level below. With emergent we mean: *Let S be a system with composition A, i.e., the various components in addition to the way they are composed. If P is a property of S, P is emergent with regard to A, if and only if no components in A possess P; otherwise P is to be regarded as a resulting property with regard to A* (Bunge, 1977: 97).

Entrepreneurial spirit. The entrepreneurial spirit may be described as follows (Roddick, 2003: 106–107):

- The vision of something new and belief in this that is so strong that belief becomes reality.
- A touch of positive madness.
- The ability to stand out from the crowd.
- Creative tension bubbling over.
- Pathological optimism.
- To act before you know!
- Basic desire for change.
- Creative energy focused on ideas, not on explicit factual knowledge.
- Being able to tell the story you want to sell.

Feedback. Giving the other person feedback, for instance with regard to their behaviour, attitudes and the like is the most important element in the area of interactive skills and emotional intelligence (Goleman, 1996; 2007). Analysis of feedback is a sure way to identify our strengths and then reinforce them (Wang, et al., 2003). Failure to give people feedback on their behaviour in some contexts may even be considered immoral.

Feed-forward. Feed-forward is regarded here as an expectation mechanism. It seems reasonable to assume that our expectations influence our behaviour in the present. It is therefore important that we make explicit to ourselves the expectations we have of a situation. By making expectations explicit, we have a greater opportunity to learn from our experiences and thus improve our performance.

Front line focus. This refers to those in the front line, i.e., in direct contact with customers, users, patients, students, etc. They have the greatest expertise, necessary information, and decision-making authority and are regarded as the most important resource in the organization because they are at the point where an organization's value creation occurs.

Global competence network. These competence networks may be divided into political, social, economic, technological and cultural patterns. It is when these five patterns interact that one may perceive the overall pattern. In the global knowledge economy it seems reasonable to assume that those who control this pattern set the conditions for economic development. These global competence networks will most likely make an impact on HR departments in companies competing for this kind of expertise in national markets.

Global competence networks are also emphasized as crucial for economic growth by OECD (2001), although they use the term *innovative clusters*. The purpose of innovative clusters and global competence networks is the development, dissemination and use of new ideas that promote wealth creation.

There is much to suggest that a greater degree of integration and cooperation between private and public sectors at the national and regional levels is an important prerequisite for initiating the innovative locomotive effect. The global competence networks are metaphorically the energy source that sustains the motion of this locomotive. It would be counterproductive to replace the locomotive once in motion. Conversely, the individual carriages of the locomotive (read: organizational level) can be changed depending on their competitive position. The individual passengers on the train create ideas and knowledge through the processes that may be called *creative chaos*. In this way we will arrive at a tripartite of the prerequisites for global competence networks. At the individual level, creative chaos occurs. At the organizational level, there will be creative destruction. At the social and global levels, creative collaboration takes place. These three processes create innovation and economic growth as an emergent, not as a *future perfectum*, i.e., a planned process with given results.

A prerequisite for the reasoning above is that tension and competition at one level require collaboration at another level. Competition and cooperation are both necessary if one is to develop innovation and economic growth, in the same manner that stability and change are necessary for flexibility. Too much of the one (stability) leads to rigidity, and too much of the other (change) leads to chaos. Understood in this way, emergents cannot be planned.

Hamel's Law of Innovation. The "law" states that only between one and two of one thousand ideas become innovations in a market (Hamel, 2002; 2012). Therefore, an infostructure must be created to ensure that ideas are continuously produced in a business.

Hidden knowledge. Hidden knowledge is what we do not know we do not know. Kirzner (1982) says that hidden knowledge is possibly the most important knowledge domain of creativity, innovation and entrepreneurship.

History's "slow fields". This refers to the fact that norms, values and actions tend to be in operation long after the functions, activities and processes that initially created them disappear, thus generating so-called *slow fields of history*. These norms, values and actions exist though they have no apparent function, contributing to maintaining a type of behaviour long after the type of behaviour is functional or meaningful.[2] For sociologists and historians it is important to determine whether norms and values have any function, or whether they are part of history's slow fields. By examining history's slow fields, it may be possible to provide better explanations for phenomena.

Implicit knowledge. This is knowledge that is spread throughout an organization but not integrated.

Informat. Intelligent robots connected with other intelligent robots in the global economy.

Information input overload. This occurs when an individual, a team, an organization or a community receives more information than they can manage to process.

In a situation characterized by information input overload the following may occur (Miller, 1978: 123):

1. Designated tasks and responsibilities are left undone.
2. Errors are made.
3. Queues of information occur.
4. Information is filtered out that should have been included.
5. Abstract formulations are made when they should have been specific.
6. Communication channels are overloaded, creating stress and tension in the system.
7. Complex situations are shunned.
8. Information is lumped together for processing.

Each of the above eight points may result in a decrease in efficiency when the system is exposed to information input overload.

Infostructure. The infostructure concerns the processes that enable the development, transfer, analysis, storage, coordination and management of data, information and knowledge. The infostructure consists of 11 generic processes, as shown in figure 8 as shown in Miller (1978). The 11 processes in the infostructure may be considered as nodes in a social network at different levels, for example team, organization, society and region, all in the global space. Together, the 11 processes comprise the totality of the infostructure.

It may be said that the *info*structure has the same importance in the knowledge society as the *infra*structure had in the industrial society.

Innovation. Innovation is here understood as any idea, practice or material element, which is perceived as new for the person using it (Zaltman et al., 1973).

Ideas are seen as the smallest unit in the innovation process (Hamel, 2002; 2012). However, this refers to the ideas that are in process of development and not fully developed ideas. Before an idea can be characterized as innovative, it must prove to be beneficial to somebody, i.e., the market must accept the idea and apply it. Consequently, the creative process of innovation is here understood as the benefit it has for a market (Amabile, 1990; Johannessen, et al., 2001: 25). Thus, it is not sufficient that an idea is new for it to be considered an innovation. An idea may have a great degree of novelty, but if it is of no benefit to anybody in the market, then it has no innovative value.

Knowledge. The definition of knowledge used here is *the systematization and structuring of information for one or more goals or purposes.*

Knowledge worker. A knowledge worker has been described by the OECD as *a person whose primary task is to generate and apply knowledge,* rather than to provide services or produce physical products (OECD, 2000a, b, c, d, e; 2001). This may be understood as a *formal definition* of a knowledge worker.

This definition does not restrict knowledge workers to creative fields, as is the case with, for example, Mosco and McKercher (2007: vii–xxiv). The OECD definition also allows for the fact that a knowledge worker may perform routine tasks. The definition also does not limit the type of work performed by knowledge workers to tasks relating to creative problem-solving strategies, unlike the definition provided by Reinhardt et al. (2011).

Knowledge enterprise. This is an enterprise that has knowledge as its most significant output. It is perhaps helpful to think of the process *input – process – output* to separate industrial enterprises from knowledge enterprises. Much knowledge and skills are needed to produce high-tech products such as computers, and there are also many knowledge workers involved in this process. However, the majority of products produced today are high-tech industrial products, and although such products require very skilled knowledge in the production process, they are nevertheless output-industrial products.

On the other hand, law firms, consulting firms and universities are examples of knowledge enterprises.

Knowledge management. The old paradigm: Management of knowledge resources in an organization. These resources may be explicit knowledge, implicit knowledge, tacit knowledge and hidden knowledge.

The new paradigm: By knowledge management we here mean governance, control and communication in social systems. Governance is here related to sustainable leadership. Control is linked to information processes and goal creation. Communication is related to the statement: who says what over which channels with which effects.

We are developing a new understanding or paradigm in knowledge management. This new paradigm is here understood as a social system perspective on knowledge processes where we consider knowledge management from an epistemological hierarchy consisting of: the philosophical perspective, the political economic perspective, the social side of knowledge management and various aspects of knowledge processes in the global economy. With this new paradigm we lift knowledge management from the more organizational understanding that the subject has had up to now, and up to a societal level.

Locomotive effect. This refers to something that generates and then reinforces an activity or development.

Modularization. An extreme fragmentation of the production process in the global knowledge economy. Production is fragmented and distributed according to the following logic: costs – quality – competence – design – innovation.

Necessary and sufficient conditions. It may often be appropriate to divide conditions or premises into *necessary conditions* and *sufficient conditions*. Necessary conditions must be present to trigger an action, but these may not be sufficient. The sufficient conditions must also be present to trigger the action.

North's action theory.[3] This action theory may be expressed in the following statement: *people act on the basis of a system of rewards as expressed in the norms, values, rules and attitudes in the culture (the institutional framework)* (North, 1990; 1993). North's action theory is also consistent with Asplund's motivation theory (refer Asplund's motivation theory).

Primary task. An organization's primary task is what the system is designed to do.

Proposition. This is an overarching hypothesis. It says something about the relationship between several variables. A proposition relates to a hypothesis in the same way the main research problem relates to research questions.

Punctuation. By punctuation (Bateson, 1972: 292–293) a distinction is drawn between cause and effect; this is done with a clear motive in mind. A causality is thus created which does not actually exist in the real world, and one is then free to discuss the effects of this cause which has been created through a process of punctuation.

A sequence of a process is selected, and then bracketed. In this way, we de-limit what is punctuated from the rest of the process. Figuratively, we may imagine this as a circle that is divided into small pieces; one piece of the circle is then selected and folded out into a straight line. This results in the creation of an artificial beginning and end. This beginning and end of course cannot exist in a circle, but only through the process of punctuation.

Social laws. Social laws constitute a pattern of a unique type. They are systemic and connected to a system of knowledge, and cannot change without the facts they represent also being changed (Bunge, 1983a; 1983b). The main differences between a statement of a law and other statements are:

1. Law statements are general.
2. Law statements are systemic, i.e., they are related to the established system of knowledge.
3. Law statements have been verified through many studies.

A pattern may be understood as variables that are stable over a specific period of time. A social law is created when an observer gains insight into the pattern. By gaining such insight, we can also predict parts of behaviour or at least develop a rough estimate within a short period of time.

Social laws are further related to specific social systems, both in time and space. However, this does not represent any objection to social laws, because this is also true of natural laws (although these have a longer time span and are of a more general nature).

Social mechanism. Robert Merton (1967) brought the notion of social mechanisms into sociology, although we can find rudiments of this in both Weber – with the Protestant ethic as an explanation for the emergence of capitalism in Europe – and Durkheim, who uses society as an explanation for a rising suicide rate. For Merton, social mechanisms are the building blocks of *middle range theories*. He defines social mechanisms as *social processes having designated consequences for designated parts of the social structure* (Merton, 1967: 43). In the 1980s and 1990s, Jon Elster developed a new notion of the role of social mechanisms in sociology (Elster, 1986; 1989). Hedstrøm and Swedberg write that, *the advancement of social theory calls for an analytical approach that systematically seeks to explicate the social mechanisms that generate and explain observed associations between events* (Hedstrøm & Swedberg, 1998: 1).

It is one thing to point out connections between phenomena. It is something quite different to point out satisfactory explanations for these relationships, which is what social mechanisms accomplish. A social mechanism tells us what will happen, how it will happen and why it will happen (Bunge, 1967). Social mechanisms are primarily analytical constructs which cannot necessarily be observed; in other words, they are epistemological, not ontological. However, social mechanisms are observable in their consequences. An intention can be a social mechanism of action. We cannot observe an intention, but we can interpret it in light of the consequences manifested through an action. Preferences can also function as a social mechanism for economic behaviour. We cannot observe a person's preferences, but we can interpret them in the light of the behavioural consequences that manifest themselves. Social mechanisms are, understood in this way, analytical constructs, indicating connections between events (Hernes, 1998).

Bunge says: "… a social mechanism is a process in a concrete system, such that it is capable of being about or preventing some change in the system as a whole or in some of its subsystems" (Bunge, 1997a: 414). By "social mechanism" here we mean those activities that promote/inhibit social processes in relation to a specific problem/phenomenon.

Material resources and technology are social mechanisms of the economic subsystem; power is a social mechanism of the political subsystem; fundamental norms and values are a social mechanism of the cultural subsystem; and human relationships are a social mechanism of the social subsystem. These system-specific social mechanisms interact with each other to achieve certain goals, maintain these systems or to avoid certain undesirable conditions in the system or the outside world.

The difficulty of discovering social mechanisms and distinguishing them from processes may be partly explained by the fact that social mechanisms are also processes (Bunge, 1997a: 414). For the application of social mechanisms, see the Boudon-Coleman diagram.

Social system. From a systemic perspective, social systems can be conceptual or concrete. Theories and analytical models are examples of conceptual systems. Further, social systems are *composed of people and their artefacts* (Bunge, 1996: 21). Social systems are held together (in systemic reasoning) by **dynamic social relations** (such as emotions, interpretations and norms) and **social actions**

(such as cooperation, solidarity, conflict and communication). None of the social actions have precedence in the systemic interpretation of social systems, such as conflict in the case of Marx, and solidarity in the case of Durkheim.

Staccato-behaviour (erratic behaviour). If organizations introduce too many change processes in succession too quickly, a phenomenon may occur called "staccato-behaviour".

If an organization does not deal with this appropriately, it seems reasonable to assume that workers will become tired, burnt-out and de-motivated. Perhaps most damaging to business, employees will lose focus on their primary task – what the business is designed to do. In addition, businesses will often experience that this leads to an increasing degree of opportunistic behaviour (Ulrich, 2013b: 260).

Systemic thinking. Systemic thinking makes a distinction between the epistemological sphere (Bunge, 1985), the ontological sphere (Bunge, 1983a), the axiological sphere (Bunge, 1989, 1996) and the ethical sphere (Bunge, 1989). Systemic thinking makes a clear distinction between intention and behaviour. Intention is something that should be *understood*, while behaviour is something that should be *explained*. To understand an intention we must study the historical factors, situations and contexts, as well as the expectation mechanisms. Behaviour must be explained with respect to the context, relationships and situation it unfolds in. What implication does the distinction between intention and behaviour have for the study of social systems?

Interpretation of meaning is an important part of the *intention aspect* in the distinction. Explanation and prediction become an essential part of the *behavioural aspect* of the distinction.

In systemic thinking it is the link between the interpretation of meaning and explanation, and prediction, which provides historical and social sciences with practical strength. By making a distinction between intention and behaviour, the historical and the social sciences are interpretive, explanatory and predictive projects. According to systemic thinking, many of the contradictions in the historical and social sciences spring from the fact that a distinction is not made between intention and behaviour. The problem of the historical and social sciences is that the actors who are studied have both intentions and they also exercise types of behaviour; however, this isn't problematic as long as we make a distinction between intention and behaviour. By simultaneously introducing the distinction between intention and behaviour, systemic thinking has made it possible to identify, for instance, partial explanations from each of two main epistemological positions, namely the naturalists and anti-naturalists (Johannessen & Olaisen, 2005; 2006), and synthesize these explanations into new knowledge.

Systemic thinking emphasizes circular causal processes, also called *interactive causal processes*, in addition to linear causal processes (Johannessen, 1996; 1997). Systemic thinking argues that to understand objective social facts, one must examine the subjective aspects of these. In systemic thinking, objective social

facts exist, but they are often more difficult to grasp than facts in the natural world, because social facts are often influenced by expectations, emotions,
prejudices, ideology and economic and social interests. "Aspect-seeing" is thus
a way of approaching these social facts.

Emergents are central to systemic thinking. A pattern behind the problem
or phenomenon is always sought in systemic investigations. Patterns may be
revealed by studying the underlying processes that constitute a phenomenon
or problem, *and the search for pattern is what scientific research is all about* (Bunge,
1996: 42).

According to systemic thinking it is a misconception to say that the facts
are social constructions. The misunderstanding involves confusing our *concepts*
concerning facts and our *hypotheses* about the facts together with the facts.
Our concepts and hypotheses are mental constructs. The facts, however, are not
mental constructs. Social need, for instance, is not a social fact; it is a mental
construct of, for instance, starvation. Starvation is a social fact. Social need is a
mental or social construction. Not being able to read is a social fact. Illiteracy is,
however, a social construction.

A *symbol* should symbolize something, just as a *concept* should delineate something.
A *hypothesis* should explain something or express something about relationships.
A conceptual *model* should say something about the relationships between
concepts. A *theory* should say something about relationships between propositions.
Physical or social facts are untouched by all these mental constructions. That one
can through constructs change social facts, or that social facts are changed as a
social consequence of using constructs, is neither original nor new.

The aim of theoretical research, according to the systemic position, is the
construction of systems, i.e., theories (Bunge, 1974: v). The order in systemic
research is thus: theory − analysis − synthesis.

In the methodological sphere, the systemic position has its main focus
on relationships, both in terms of concrete things, ideas and knowledge.
Consequently, systemic thinking encourages interdisciplinary and multidisciplinary approaches to problems or phenomena.

The systemic position thus attempts to bridge the gap between methodological individualism and methodological collectivism, which is considered the
classic controversy in historical and social sciences.

The perceptions that an observer has about social systems will influence
his/her actions, regardless of whether the perceptions are true or fallacious.
Systemic investigations start, therefore, writes Bunge *from individuals embedded
in a society that preexists them and watch how their actions affect society and alter it*
(Bunge, 1996: 241). The study of social systems from a systemic perspective for
these reasons always includes the triad: actors − observers − social systems.

The observer tries to uncover a system's composition, environment and
structure. Then the actors' subjective perception of composition, environment
and structure are examined. In other words, both the subjective and objective
aspects are studied. When we wish to study changes in social systems, from a
systemic point of view, we have to examine the social mechanisms (drivers) that

influence changes; both internal and external social mechanisms must be identified. This study takes place within the four subsystems: the economic, political,
cultural and relational. According to systemic thinking, social changes occur
along seven axes:

1. As an *expectation* of new relationships, values, power constellations, technologies and distribution of material resources.
2. As a result of our *beliefs* (mental models) about relationships, values, power
   constellations, technical and material resources.
3. As a result of *psychological elements* such as irritation, crisis, discomfort,
   unsatisfactory life, unworthy life and loss of well-being.
4. As a result of *communication* in and between systems.
5. As a result of an *understanding of connections* (contextual understanding).
6. As a result of learning and new *self-knowledge*.
7. As a result of *new ideas* and ways of thinking.

Historiography, from a systemic perspective, has one clear goal: to investigate
what happened, where it happened, when it happened, how it happened, why
it happened and with what results.

Systemic assumptions related to historiography and social sciences may be
expressed in the following (Bunge 1998: 263):

a. The past has existed.
b. Parts of the past can be known.
c. Every uncovering of the past will be incomplete.
d. New data, techniques, and systemizations and structuring will reveal new
   aspects of the past.
e. Historical knowledge is developed through new data, discoveries, hypotheses and approaches.

In systemic thinking if changes are to take place, then the material will sometimes be given precedence; at other times, ideology, ideas and thinking are given
precedence. In other contexts, there is a systemic link between the material and
ideas that is needed to bring about changes. In such contexts, it is difficult and
irrelevant to say what is the primary driver, i.e., the material or ideas; this would
be on par with discussing what came first, the chicken or the egg.

The processes that drive social change, according to a systemic perspective, are the interaction between the economic, political, relational and cultural subsystems. In some situations, one of these four perspectives will prevail,
whereas in others it will be one or more of the four subsystems that are the
drivers of social change. In many cases, it is precisely the interaction between
the four subsystems that leads to social changes.

In this context the systemic perspective may be described by saying that
material conditions/energy, such as economic relationships, may provide the
ground from which ideologies develop, but that these ideologies in return

influence the development of the material. Whether material conditions/ energy or ideology comes first is often determined by a historiographical punctuation process (Bateson, 1972: 163).

The systemic perspective balances historical materialism and historical idealism. It assumes that overall social changes are the result of economic, political, social and cultural factors, in addition to the interaction between material conditions/energy and ideas. Furthermore, a systemic perspective views any society as being interwoven into its surroundings (Bunge, 1998: 275). When a historian considers a historical situation – such as the massacre in Van in April 1915 – from this perspective then he is trying *to throw light upon the internal working of a past culture and society* (Stone, 1979: 19).

The systemic position attempts to view the relevant event in a larger context, in order to find *the patterns which combine* (Bateson, 1972: 273–274), because *change depends upon feedback loop* (Bateson, 1972: 274). Bunge says about this position: *by placing the particular in a sequence, adopting a broad perspective the systemist overcomes the idiographic/nomothetic duality, …, as well as the concomitant narrative/ structural opposition* (Bunge 1998: 275). This means, metaphorically, that the systemic researcher uses a microscope, telescope and a helicopter to investigate patterns over time.

Systemic research strategy is a *zig-zagging between the micro-meso and macro-levels* (Bunge, 1998: 277). Through a systemic research strategy the researcher has ample opportunities to use a Boudon-Coleman diagram.

Systemic thinking examines four types of changes[4].

Type I change concerns individuals who change history such as Genghis Khan, Hitler, Stalin and Mao Zedong

Type II change concerns groups of people acting together who change history. Examples of Type II change include the invasion of the Roman Empire by peoples from the north; and the Ottoman expansion into the Balkans between the late 1400s and when the Ottoman Empire was pushed back partly due to nationalist liberation movements in the early 1900s.

Type III change includes changes in history that are caused by natural disasters such as the volcanic eruption that destroyed Pompeii. Climate change may also be said to be an example of a type III change.

Type IV change involves a total change in the way of thinking such as the emergence of new religions, like Islam, or a new political ideology such as Marxism.

The systemic researcher attempts to explore the relationship between the four types of changes. A single event is in itself not necessarily of special interest to the systemic researcher; rather, the focus is on the *system of events* of which the single event is a part.

All the social sciences are used in the systemic position to seek insight, understanding and explain a phenomenon or problem.

Tacit knowledge. Knowledge that is difficult to communicate to others as information. It is also very difficult, if at all possible, to digitize.

Technology. Technology, in systemic thought, is the scientific study of artefacts (Bunge, 1985: 219–231). Artefacts may be classified as instruments, machines, automats and informats.

The knowledge-based perspective. The knowledge-based perspective is defined here as creating, expanding and modifying internal and external competencies to promote what the organization is designed to do (Grant, 2003: 203).

Theory. Here understood as a system of propositions (Bunge, 1974: v).

Notes

1 Asplund's motivation theory, a term we use here, is based on Asplund's research.
2 Asplund (1970: 55) refers to a similar phenomenon when he discusses Simmel. He points out that the norms that may have had a positive function during a historic phase become in a later phase dysfunctional.
3 North's action theory is a term we use here based on North's research.
4 The four types of changes are related to Bateson's (1972: 279–309) work on different types of learning, especially those discussed in his article *Logical types of learning and communication.*

Bibliography

Amabile, T. (1988). A model of creativity and innovation in Organizations, in Staw, B.M. & Cummings, L.L. (eds.). Research in organizational behavior, 9: 123–167.
Amabile, T. (1996). Creativity in context, Westview Press, New York.
Armstrong, M. (2014a). Armstrong's handbook of strategic human resource management, Kogan Page, New York.
Armstrong, M. (2014b). Armstrong's handbook of human resource management practice, Kogan Page, New York.
Asplund, J. (1970). Om undran innfør samhället, Argos, Stockholm.
Asplund, J. (2010). Om undran innfør samhället, Argos, Stockholm.
Bateson, G. (1972). Steps to an ecology of mind, Intertext Books, London.
Bleuer, H.; Bouri, M. & Mandada, F.C. (2017). New trends in medical and service robots, Springer, London.
Bleuer, H. & Bouri, M. (2017). New trends in medical and service robots: Assistive, surgical and educational robotics, Springer, London.
Boudon, R. (1981). The logic of social action, Routledge, London.
Boxall, P.F. & Purcell, J. (2003). Strategy and human resource management, Palgrave Macmillan, Basingstoke.
Boxall, P.F. & Purcell, J. (2010). An HRM perspective on employee participation, in Wilkinson, A.; Golan, P.J.; Marchington, M. & Lewins, D. (eds.). The Oxford handbook of participation in organizations, Oxford University Press, Oxford, s. 129–151.
Boxall, P.F.; Purcell, J. & Wright, P. (2007). Human resource management: Scope, analysis, and significance, in Boxall, P.F.; Purcell, J. & Wright, P. (eds.). The Oxford handbook of human resource management, Oxford University Press, Oxford. s. 1–16.

Brockbank, W. (2013). Overview and logic, in Ulrich, D.; Brockbank, W.; Younger, J. & Ulrich, M. (eds.). Global HR competencies: Mastering competitive value from the outside in, McGraw Hill, New York. s. 3–27.

Bunge, M. (1967). Scientific research, Vol. 3, in studies of the foundations methodology and philosophy of science, Springer Verlag, Berlin.

Bunge, M. (1974). Sense and reference, Reidel, Dordrecht.

Bunge, M. (1977). Treatise on basic philosophy. Vol. 3. Ontology I: The furniture of the world, D. Reidel, Dordrecht, Holland.

Bunge, M. (1979). A world of systems, Reidel, Dordrecht.

Bunge, M. (1983a). Exploring the world: Epistemology & methodology I, Reidel, Dordrecht.

Bunge, M. (1983b). Understanding the world: Epistemology & methodology II, Reidel, Dordrecht.

Bunge, M. (1985). Philosophy of science and technology. Part I: Epistemology & methodology III, Reidel, Dordrecht.

Bunge, M. (1989). Ethics: The good and the right, Reidel, Dordrecht.

Bunge, M. (1996). Finding philosophy in social science, Yale University Press, New Haven.

Bunge, M. (1997a). Mechanism and explanation. Philosophy of the Social Sciences 27: 410–465.

Bunge, M. (1997b). Foundations of biophilosophy, Springer Verlag, Berlin.

Bunge, M. (1998). Philosophy of science: From problem to theory, Volume one, Transaction Publishers, New Jersey.

Coleman, J.S. (1990). Foundations of social theory, Harvard University Press, Belknap Press, Cambridge, MA.

Duncan, R. (1976). The ambidextrous organization: Designing dual structures for innovation, in Kilman, R.H.; Pondy, L.R. & Slevin, D. (eds.). The management of organization, North Holland, New York. s. 167–188.

Elster, J. (1986). Rational choice, New York University Press, New York.

Elster, J. (1989). Nuts and bolts for the social sciences, Cambridge University Press, Cambridge.

Goleman, D. (1996). Emotional intelligence, Bloomsbury Publishing, New York.

Goleman, D. (2007). Social intelligence, Arrow books, New York.

Grant, R.M. (2003). The knowledge-based view of the firm, in Faulkner, D. & Campbell, A. (red.). The Oxford handbook of strategy, Oxford University Press, Oxford. s. 203–231.

Hamel, G. (2002). Leading the revolution: How to thrive in turbulent times by making innovation a way of life, Harvard Business School Press, Boston.

Hamel, G. (2012). What matters now: How to win in a world of relentless change ferocious competition, and unstoppable innovation, John Wiley & Sons, New York.

Hamel, G. & Prahalad, C.K. (1996). Competing for the future, Harvard Business School Press, Boston.

Hedstrøm, P. & S.R. Swedberg (1998). Social mechanisms: An introductory essay, in Hedstrøm, P. & R. Swedberg (red.). Social mechanisms: An analytical approach to social theory, Cambridge University Press, Cambridge. pp. 20–32.

Helfat, C.E.; Finkelstein, S.; Mitchell, W.; Peteraf, M.A.; Singh, H.; Teece, D.J. & Winter, S.G. (2007). Dynamic capabilities: Understanding strategic change in organizations, Blackwell, Oxford.

Hernes, G. (1998). Real virtuality, in Peter Hedstrøm and Richard Swedberg (eds.). Social mechanisms: An analytical approach to social theory, Cambridge University Press, Cambridge. pp. 74–102.

Hewing, M. (2013). Collaboration with potential users for discontinuous innovation, Springer Gabler, Potsdam.

Johannessen, J-A. (1996). Systemics applied to the study of organizational fields: Developing systemic research strategy for organizational fields, Kybernetes, vol. 25, 1: 33–51.

Johannessen, J-A. (1997). Aspects of ethics in systemic thinking" Kybernetes, 26, 9: 983–1001.

Johannessen, J-A.; Olaisen, J. & Olsen, B. (2001). Mismanagement of tacit knowledge: The importance of tacit knowledge, the danger of information technology, and what to do about it? International Journal of Information Management, 21, 3: 3–20.

Johannessen, J.-A. & J. Olaisen (2005). Systemic philosophy and the philosophy of social science-Part I: Transcendence of the naturalistic and the anti-naturalistic position in the philosophy of social science" I Kybernetes, 34, 7/8: 1261–1277.

Johannessen, J.-A. & J. Olaisen (2006). Systemic philosophy and the philosophy of social science-Part II: The systemic position" I Kybernetes 34, 9/10: 1570–1586.

Kirzner, S. (1982). The theory of entrepreneurship in economic growth. I Kent, C.A.; Sexton, D.L. & Vesper, K.H. (red.). Encyclopedia of entrepreneurship, Prentice Hall, Englewood Cliffs. NJ.

Merton, R.K. (1967). Social theory and social structure, Free Press, London.

Mosco, V. & McKercher, C. (2007). Introduction: Theorizing knowledge labor and the information society. Knowledge workers in the information society, Lexington Books, Lanham.

Miller, J.G. (1978). Living systems, McGraw-Hill, New York.

Nelson, R.R. & Winter, S.G. (1982). An evolutionary theory of economic change, Harvard University Press, Cambridge, MA.

North, D.C. (1990). Institutions, institutional change and economic performance, Cambridge University Press, Cambridge, MA.

North, D. (1993). Nobel lecture: www.nobelprize.org/nobel_prizes/economics/laureates/1993/ north-lecture.html#not2, lesedato, 4.5.2012.

OECD (2000a), A new economy? The changing role of innovation and information technology in growth, Paris.

OECD (2000b), Economic outlook, Paris.

OECD (2000c), Education at a glance: OECD indicators, CERI, Paris.

OECD (2000d), "ICT skills and employment, working party on the information economy", Paris, 15 November, DSTI/ICCP/IE (2000)7.

OECD (2000e), Knowledge management in the learning society, CERI, Paris.

OECD (2001). Innovative clusters: Driving of national innovation-systems, OECD, Paris.

O'Reilly, C.A. & Tushman, M.L. (2004). The ambidextrous organization, Harvard Business Review, 82, 4: 74–81.

O'Reilly, C.A. & Tushman, M.L. (2007). Ambidexterity as a dynamic capability: Resolving the innovators dilemma, Harvard Business School Press, Boston.

O'Reilly, C.A. & Tushman, M.L. (2011). Organizational ambidexterity in action: How managers explore and exploit, California Management Review, 53, 4: 5–22.

Reinhardt, W., Smith, B.; Sloep, P. & Drachler, H. (2011). Knowledge worker roles and actions – Results of two empirical studies, Knowledge and Process Management 18 3: 150–174.

Roddick, D.A. (2003). The grassroots entrepreneur, Elbæk, U. (ed.) Kaospilot A-Z, Narayana Press, Gylling. pp. 34–45.

Stone, J. (1979). The revival of narrative: Reflections on a new old history, Past and Present, 85: 3–24.

Storey, J.; Ulrich, D. & Wright, P.M. (2009). Introduction, in Storey, J.; Wright, P.M. & Ulrich, D. (eds.). The Routledge companion to strategic human resource management, Routledge, London. pp. 3–15.

Teece, D.J. (2013). Dynamic capabilities and strategic management: Organizing for innovation, OUP, Oxford.

Teece, D.; Pisano; G. & Shuen, A. (1997). Dynamic capabilities and strategic management, Strategic Management Journal, 18, 7: 509–533.

Thota, H. & Munir, Z. (2011). Key concepts in innovation, Palgrave Macmillan, London.

Tushman, M.L. & O'Reilly, C.A. (1996). Ambidextrous organizations: Managing evolutionary and revolutionary change, California Management Review 38, 4: 8–30.

Ulrich, D. (2013a). Foreword, in Ulrich, D.; Brockbank, W.; Younger, J. & Ulrich, M. (eds.). Global HR competencies: Mastering competitive value from the outside in, McGraw Hill, New York. s. v–xxi.

Ulrich, D. (2013b). Future of global HR: What's next?, in Ulrich, D.; Brockbank, W.; Younger, J. & Ulrich, M. (eds.). Global HR competencies: Mastering competitive value from the outside in, McGraw Hill, New York. pp. 255–268.

Ulrich, D. & Brockbank, W. (2005). The HR value proposition, Harvard Business School Press, Boston, MA.

Vadakkepat, P. & Goswami P. (eds.). (2018). Humanoid robotics: A reference, Springer, London.

Wang, Q-G.; Lee, T.H. & Lin, C. (2003). Relay feedback: Analysis, identification and control, Springer, London.

Wernerfelt, B. (1984). A resource-based view of the firm, Strategic Management Journal, 5, 2: 171–180.

White, J. & Younger, J. (2013). The global perspective, in Ulrich, D.; Brockbank, W.; Younger, J. & Ulrich, M. (eds.); Global HR competencies: Mastering competitive value from the outside in, McGraw Hill, New York. pp. 27–53.

Wilson, M. (2017). Implementation of robot systems, Butterworth-Heinemann, New York.

Winfield, A. (2012). Robotics, OUP, Oxford.

Wright, P.M.; Boudreau, J.W.; Pace, D.A.; Libby Sartain, E.; McKinnon, P.; Antoine, R.L. (Eds.). (2011). The chief HR officer: Defining the new role of human resource leaders, Jossey-Bass, London.

Wright, P.; Dunford, B. & Snell, S. (2001). Human resources and the resource based view of the firm, Journal of Management, 27: 701–721.

Winter, S.G. (2003). Understanding dynamic capabilities, Strategic Management Journal, 24: 991–995.

Zaltman, G., Duncan, R., & Holbeck, J. (1973). Innovations and organizations. New York: Wiley.

Appendix 2: Systemic thinking

Systemic thinking makes a distinction between the epistemological sphere (Bunge, 1985), the ontological sphere (Bunge, 1983a), the axiological sphere (Bunge, 2003) and the ethical sphere (Bunge, 1989). Systemic thinking makes a clear distinction between intention and behaviour. Intention is something that should be *understood*, while behaviour is something that should be *explained*. To understand an intention we must study the historical factors, situations and contexts, as well as the expectation mechanisms. Behaviour must be explained with respect to the context, relationships and situation it unfolds in. What implication does the distinction between intention and behaviour have for the study of social systems?

Interpretation of meaning is an important part of the *intention aspect* in the distinction. Explanation and prediction become an essential part of the *behavioural aspect* of the distinction.

In systemic thinking it is the link between the interpretation of meaning and explanation, and prediction, which provides historical and social sciences with practical strength. By making a distinction between intention and behaviour, the historical and the social sciences are interpretive, explanatory and predictive projects. According to systemic thinking, many of the contradictions in the historical and social sciences spring from the fact that a distinction is not made between intention and behaviour. The problem of the historical and social sciences is that the actors who are studied have both intentions and they also exercise types of behaviour; however, this isn't problematic as long as we make a distinction between intention and behaviour. By simultaneously introducing the distinction between intention and behaviour, systemic thinking has made it possible to identify, for instance, partial explanations from each of two main epistemological positions, namely, the naturalists and anti-naturalists, and synthesize these explanations into new knowledge (Johannessen & Olaisen, 2005; 2006).

Systemic thinking emphasizes circular causal processes, also called *interactive causal processes*, in addition to linear causal processes (Johannessen, 1995). Systemic thinking argues that to understand objective social facts, one must examine the subjective aspects of these. In systemic thinking, objective social facts exist, but they are often more difficult to grasp than facts in the natural world, because social facts are often influenced by expectations, emotions, prejudices, ideology and economic and social interests. "Aspect-seeing" is thus a way of approaching these social facts.

Emergents are central to systemic thinking. An emergent occurs if something new turns up on one level that has not previously existed on the level below. With emergent we mean: *Let S be a system with composition A, i.e., the various components in addition to the way they are composed. If P is a property of S, P is emergent with regard to A, if and only if no components in A possess P; otherwise P is to be regarded as a resulting property with regard to A* (Bunge, 1977: 97).

A pattern behind the problem or phenomenon is always sought in systemic investigations. Patterns may be revealed by studying the underlying processes that constitute a phenomenon or problem, *and the search for pattern is what scientific research is all about* (Bunge, 1996: 42).

According to systemic thinking it is a misconception to say that the facts are social constructions. The misunderstanding involves confusing our *concepts* concerning facts and our *hypotheses* about the facts together with the facts. Our concepts and hypotheses are mental constructs. The facts, however, are not mental constructs. Social need, for instance, is not a social fact; it is a mental construct of, for instance, starvation. Starvation is a social fact. Social need is a mental or social construction. Not being able to read is a social fact. Illiteracy is, however, a social construction.

A *symbol* should symbolize something, just as a *concept* should delineate something. A *hypothesis* should explain something or express something about relationships. A conceptual *model* should say something about the relationships between concepts. A *theory* should say something about relationships between propositions. Theory is in systemic thinking understood as a system of propositions (Bunge, 1974: v).

Physical or social facts are untouched by all these mental constructions. That one can through constructs change social facts, or that social facts are changed as a social consequence of using constructs, is neither original nor new.

The aim of theoretical research, according to the systemic position, is the construction of systems, i.e., theories (Bunge, 1974: v). The order in systemic research is thus: theory – analysis – synthesis.

In the methodological sphere, the systemic position has its main focus on relationships, both in terms of concrete things, ideas and knowledge. Consequently, systemic thinking encourages interdisciplinary and multidisciplinary approaches to problems or phenomena.

The systemic position thus attempts to bridge the gap between methodological individualism and methodological collectivism, which is considered the classic controversy in historical and social sciences. In methodology systemic thinking is also focused on the Boudon-Coleman diagram. This research methodology was developed by Mario Bunge (Bunge, 1979) based on insights made by the sociologists Boudon and Coleman. The purpose of the diagram is to show the relationship between the various levels such as the macro- and micro-levels. For instance, it is shown how changes at the macro-level, such as technological innovations in feudal society, can lead to increased income at the micro-level. However, it was shown that technological innovations could lead to weakening of the semi-feudal structures because dependency on landowners was reduced. Consequently, the landowners opposed such changes especially in the case of technological innovations, which Boudon has shown in his research (Boudon, 1981: 100). Coleman (Coleman, 1990: 7–12) started at the macro-level, went to the individual level to find explanations and finally ended up at the macro-level again.

An important purpose of Bunge's Boudon-Coleman diagram is to identify social mechanisms that maintain or change the phenomenon or problem under investigation (as mentioned above, in Boudon's analysis of semi-feudal society). Bunge's Boudon-Coleman diagram may be said to represent a "mixed strategy"; Bunge says the following: *when studying systems of any kind (a) reduce them to their components (at some level) and the interaction among these, as well as among them and environmental items, but acknowledge and explain emergence* (see the chapter on concepts) *whenever it occurs, and (b) approach systems from all pertinent sides and on all relevant levels, integrating theories or even research fields whenever unidisciplinarity proves to be insufficient* (Bunge, 1998: 78). The purpose of this research strategy is to arrive at a deeper and more complete explanation of a system's behaviour.

It was Robert Merton (1967) who brought the notion of social mechanisms into sociology, although we can find rudiments of this in both Weber – with the Protestant ethic as an explanation for the emergence of capitalism in Europe – and Durkheim, who uses society as an explanation for a rising suicide rate. For Merton, social mechanisms are the building blocks of *middle range theories*. He defines social mechanisms as *social processes having designated consequences for designated parts of the social structure* (Merton, 1967: 43). In the 1980s and 1990s, Jon Elster developed a new notion of the role of social mechanisms in sociology (Elster, 1986; 1989). Hedstrom and Swedberg write that, *the advancement of social theory calls for an analytical approach that systematically seeks to explicate the social mechanisms that generate and explain observed associations between events* (Hedstrøm & Swedberg, 1998: 1).

It is one thing to point out connections between phenomena. It is something quite different to point out satisfactory explanations for these relationships, which is what social mechanisms accomplish. A social mechanism tells us what will happen, how it will happen and why it will happen (Bunge, 1967). Social mechanisms are primarily analytical constructs which cannot necessarily be observed; in other words, they are epistemological, not ontological. However, social mechanisms are observable in their consequences. An intention can be a social mechanism of action. We cannot observe an intention, but we can interpret it in light of the consequences manifested through an action. Preferences can also function as a social mechanism for economic behaviour. We cannot observe a person's preferences, but we can interpret them in the light of the behavioural consequences that manifest themselves. Social mechanisms are, understood in this way, analytical constructs, indicating connections between events (Hernes, 1998).

Bunge says: "… a social mechanism is a process in a concrete system, such that it is capable of being about or preventing some change in the system as a whole or in some of its subsystems" (Bunge, 1997: 414). By "social mechanism" here we mean those activities that promote/inhibit social processes in relation to a specific problem/phenomenon.

Material resources and technology are social mechanisms of the economic subsystem; power is a social mechanism of the political subsystem; fundamental norms and values are a social mechanism of the cultural subsystem; and human relationships are a social mechanism of the social subsystem. These system-specific social mechanisms interact with each other to achieve certain goals, maintain these systems or to avoid certain undesirable conditions in the system or the outside world.

The difficulty of discovering social mechanisms and distinguishing them from processes may be partly explained by the fact that social mechanisms are also processes (Bunge, 1997: 414).

The perceptions that an observer has about social systems will influence his/her actions, regardless of whether the perceptions are true or fallacious. Systemic investigations start, therefore, writes Bunge *from individuals embedded in a society that preexists them and watch how their actions affect society and alter it* (Bunge, 1996: 241). The study of social systems from a systemic perspective for these reasons always includes the triad: actors – observers – social systems.

From a systemic perspective, social systems can be conceptual or concrete. Theories and analytical models are examples of conceptual systems. Further, social systems are *composed of people and their artefacts* (Bunge, 1996: 21). Social systems are held together (in systemic reasoning) by **dynamic social relations** (such as emotions, interpretations and norms) and **social actions** (such as cooperation, solidarity, conflict and communication). None of the social actions have precedence in the systemic interpretation of social systems, such as conflict in the case of Marx, and solidarity in the case of Durkheim.

The observer tries to uncover a system's composition, environment and structure. Then the actors' subjective perception of composition, environment and structure are examined. In other words, both the subjective and objective aspects are studied. When we wish to study changes in social systems, from a systemic point of view, we have to examine the social mechanisms (drivers) that influence changes; both internal and external social mechanisms must be identified. This study takes place within the four subsystems: the economic, political, cultural and relational. According to systemic thinking, social changes occur along seven axes:

1. As an *expectation* of new relationships, values, power constellations, technologies and distribution of material resources.
2. As a result of our *beliefs* (mental models) about relationships, values, power constellations, technical and material resources.
3. As a result of *psychological elements* such as irritation, crisis, discomfort, unsatisfactory life, unworthy life and loss of well-being.

4. As a result of *communication* in and between systems.
5. As a result of an *understanding of connections* (contextual understanding).
6. As a result of learning and new *self-knowledge*.
7. As a result of *new ideas* and ways of thinking.

Historiography, from a systemic perspective, has one clear goal: to investigate what happened, where it happened, when it happened, how it happened, why it happened and with what results.

Systemic assumptions related to historiography and social sciences may be expressed in the following (Bunge 1998: 263):

a. The past has existed.
b. Parts of the past can be known.
c. Every uncovering of the past will be incomplete.
d. New data, techniques, and systemizations and structuring will reveal new aspects of the past.
e. Historical knowledge is developed through new data, discoveries, hypotheses and approaches.

In systemic thinking if changes are to take place, then the material will sometimes be given precedence; at other times, ideology, ideas and thinking are given precedence. In other contexts, there is a systemic link between the material and ideas that is needed to bring about changes. In such contexts, it is difficult and irrelevant to say what is the primary driver, i.e., the material or ideas; this would be on par with discussing what came first, the chicken or the egg.

The processes that drive social change, according to a systemic perspective, are the interaction between the economic, political, relational and cultural subsystems. In some situations, one of these four perspectives will prevail, whereas in others it will be one or more of the four subsystems that are the drivers of social change. In many cases, it is precisely the interaction between the four subsystems that leads to social changes.

In this context the systemic perspective may be described by saying that material conditions/energy, such as economic relationships, may provide the ground from which ideologies develop, but that these ideologies in return influence the development of the material. Whether material conditions/energy or ideology comes first is often determined by a historiographical punctuation process (Bateson, 1972: 163).

The systemic perspective balances historical materialism and historical idealism. It assumes that overall social changes are the result of economic, political, social and cultural factors, in addition to the interaction between material conditions/energy and ideas. Furthermore, a systemic perspective views any society as being interwoven into its surroundings (Bunge, 1998: 275). When a historian considers a historical situation – such as the massacre in Van in April 1915 – from this perspective then he is trying *to throw light upon the internal working of a past culture and society* (Stone, 1979: 19).

The systemic position attempts to view the relevant event in a larger context, in order to find *the patterns which combine* (Bateson, 1972: 273–274), because *change depends upon feedback loop* (Bateson, 1972: 274). Bunge says about this position: *by placing the particular in a sequence, adopting a broad perspective the systemist overcomes the idiographic/nomothetic duality, …, as well as the concomitant narrative/ structural opposition* (Bunge 1998: 275). This means, metaphorically, that the systemic researcher uses a microscope, telescope and a helicopter to investigate patterns over time.

Systemic research strategy is a *zig-zagging between the micro-meso and macro-levels* (Bunge, 1998: 277). Through a systemic research strategy the researcher has ample opportunities to use a Boudon–Coleman diagram.

Systemic thinking examines four types of changes[1].

Type I change concerns individuals who change history such as Genghis Khan, Hitler, Stalin and Mao Zedong.

Type II change concerns groups of people acting together who change history. Examples of Type II change include the invasion of the Roman Empire by peoples from the north; and the Ottoman expansion into the Balkans between the late 1400s and when the Ottoman Empire was pushed back partly due to nationalist liberation movements in the early 1900s.

Type III change includes changes in history that are caused by natural disasters such as the volcanic eruption that destroyed Pompeii. Climate change may also be said to be an example of a type III change.

Type IV change involves a total change in the way of thinking such as the emergence of new religions, like Islam, or a new political ideology such as Marxism.

The systemic researcher attempts to explore the relationship between the four types of changes. A single event is in itself not necessarily of special interest to the systemic researcher; rather, the focus is on the *system of events* of which the single event is a part.

All the social sciences are used in the systemic position to seek insight, understanding and explain a phenomenon or problem.

Note

1 The four types of changes are related to Bateson's (1972: 279–309) work on different types of learning, especially those discussed in his article *Logical types of learning and communication*.

Bibliography

Bateson, G. (1972). Steps to an ecology of mind, Intex Books, London.
Boudon, R. (1981). The logic of social action, Routledge, London.
Bunge, M. (1967). Scientific research, Vol. 3, in studies of the foundations methodology and philosophy of science, Springer Verlag, Berlin.
Bunge, M. (1974). Sense and reference, Reidel, Dordrecht.

Bunge, M. (1977). The furniture of the world. Reidel, Dordrecht.

Bunge, M. (1979). A world of systems, Reidel, Dordrecht.

Bunge, M. (1983a). Exploring the world, Reidel, Dordrecht.

Bunge, M. (1983b). Understanding the world, Reidel, Dordrecht.

Bunge, M. (1985). Philosophy of science and technology, Part I, Reidel, Dordrecht.

Bunge, M. (1989). Ethics: The good and the right, Reidel, Dordrecht.

Bunge, M. (1996). Finding philosophy in social science. New Haven: Yale University Press.

Bunge, M. (1997). Mechanism and explanation. Philosophy of the Social Sciences 27: 410–465.

Bunge, M. (1998). Philosophy of science: From problem to theory, Volume one, Transaction Publishers, New Jersey.

Bunge, M. (2003). Emergence and convergence, University of Toronto Press, Toronto.

Bunge, M. (2015). Political philosophy, Transaction Publisher, London

Collingwood, R.G. (1946). The idea of history, The University Press, Oxford.

Coleman, J.S. (1990). Foundations of social theory, Harvard University Press, Belknap Press, Cambridge, MA.

Elster, J. (1986). Rational choice, New York University Press, New York.

Elster, J. (1989). Nuts and bolts for the social sciences, Cambridge University Press, Cambridge.

Hedstrøm, P. & S.R. Swedberg (1998). Social mechanisms: An introductory essay, in Hedstrøm, P. & Swedberg R. (red.). Social mechanisms: An analytical approach to social theory, Cambridge University Press, Cambridge. pp. 20–32.

Hernes, G. (1998). Real virtuality, in Peter Hedstrøm and Richard Swedberg (eds.). Social mechanisms: An analytical approach to social theory. Cambridge University Press, Cambridge. pp. 74–102.

Johannessen, J-A. (1995). Systemics applied to the study of organizational fields: Developing a systemic research strategy for organizational fields, Kybernetes, høst 1995.

Johannessen, J-A. (1996). Systemics applied to the study of organizational fields: Developing systemic research strategy for organizational fields, Kybernetes, vol. 25, 1: 33–51.

Johannessen, J-A. (1997). Aspects of ethics in systemic thinking" i Kybernetes, 26, 9: 983–1001.

Johannessen, J.-A. & J. Olaisen (2005). Systemic philosophy and the philosophy of social science-Part I: Transcendence of the naturalistic and the anti-naturalistic position in the philosophy of social science" I Kybernetes 34, 7/8, 1261–1277.

Johannessen, J.-A. & J. Olaisen (2006). Systemic philosophy and the philosophy of social science-Part II: The systemic position" I Kybernetes 34, 9/10, 1570–1586.

Merton, R.K. (1967). Social theory and social structure, Free Press, London.

Stone, J. (1979). The revival of narrative: Reflections on a new old history, Past and Present, 85: 3–24.

Turner, J.H. (1987). Analytical theorizing, I Giddens, A. & Turner J.H. (red.). Social theory today, Polity Press, New York, s. 156–195.

Turner, J.H. (1988). A theory of social interaction, Polity Press, New York.

Turner, J.H. (1991). The structure of sociological theory, Wadsworth Publishing Company, Belmont, CA.

Index

for 67; reflective synthetic ethics 68–69; requisite variety 62

TUG 78
Turing, Alan 9, 37
Turing test 3, 9, 37–38

United Nations (UN) 89

"vacuum cleaner" robots 31, 35
Vietnam War 12, 13

"Watson" computing system 9
white lies 27, 30
World Health Organization 89

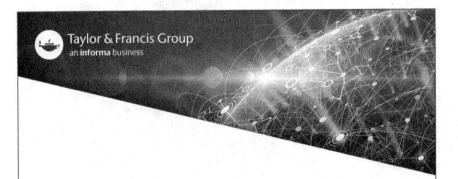

Printed in the United States
by Baker & Taylor Publisher Services